NOBODY
CARES

FRANK SAFETYMAN BURG

NOBODY CARES

THE STORY OF THE WORLD FROM SAFETYMAN

Archway Publishing books may be ordered through booksellers or by contacting:

Archway Publishing
1663 Liberty Drive
Bloomington, IN 47403
www.archwaypublishing.com
1 (888) 242-5904

ISBN: 978-1-4808-9101-2 (sc)
ISBN: 978-1-4808-9099-2 (hc)
ISBN: 978-1-4808-9100-5 (e)

Library of Congress Control Number: 2020911974

Print information available on the last page.

Archway Publishing rev. date: 6/30/2020

To my wife,
Marcella Burg.
Without her, none of this would have been possible.
She lets me do my thing!

CONTENTS

INTRODUCTION

SAFETY AND HEALTH MAY BE ONE OF THE MOST UNDERESTI-mated and underappreciated fields of study and work in our society. I have been spending my life learning that so many fields of study are linked and connected for our protection and survival. There is nothing that can be more important than our protection and survival. I could see that the work I was doing for the improvement of the safety and security of our world required knowledge from virtually every source available. It was computers and the age of information that made my work possible. The study of safety and health and the prevention of accidents (I like to call them incidents) and illnesses was more comprehensive than other fields. Protecting our lives and our quality of life is so essential, and as the years passed, I found I needed additional tools to do my job. I needed mathematics, engineering, psychology, physiology, ergonomics, and specialized information on virtually every job and trade. Statistics and data became especially important. I collected a body of knowledge that few individuals have the possibility of learning. I needed to share what I have learned with other people because the information became so vital to survival. The excitement of my experience and the epiphany of its value and importance made me a warrior to enable more people to study and learn about the field of safety and health. The information must be shared as a tool for each of us to exist in a hostile world.

These are the stories of the life of the safetyman. It is a book about the field and the science of safety, but it also is a book about the life, trials, and tribulations of a person who has been dedicated to safety and health. Every single person living in our complex and dangerous world must be a safetyman or a safety woman to avoid obstacles and dangers that will be life changing. We are all up against danger every single day, perhaps every single moment, and the idea is that the information and examples provided from my experience can be the difference between life and death. The simplest task of getting your mail, using a stool to change an energized lightbulb, or pulling leaves out of your gutter can put your life, family, and your financial security in jeopardy. Reading this book should make you think about what you are going to do before you act.

Many people believe that they are what we call bulletproof. This is a common phrase in the safety profession. They have yet to take a bullet and don't expect to take a bullet. Because they have never sustained a serious injury, they believe it will never happen to them, or even if it does happen to them, they believe that they will just heal and come back as good as ever. It is only through years of living and experience that we learn that a single solitary event, a second of carelessness, or just being at the wrong place at the wrong time can have life-changing consequences.

The safety and health problems are complicated, involving the complex interaction between the physical world, acts of men and women, and human behavior. This interaction requires study and

evaluation from the perspective of understanding the potential hazards and evaluating and testing feasible and reasonable interventions.
You must take care of yourself, but you also need professionals to
develop and design products and services that keep your safety and
health in mind. Many people think that looking out for yourself will
solve the problems, but everywhere you go, there are situations where
some professional engineer, architect, or service person is thinking
about safety or ignoring it. Each day that you are alive, you are at
serious risk, as are the people you love. You will need to take action
to plan and organize your activities to keep yourself and the ones you
love alive and protected. This book is a guide to survival in a hostile
world of hazards.

ANXIETY

NOTICE THE FEARFUL LOOK IN THE EYES OF ANIMALS THAT must endure predators and hazards in our hostile world. Roadkill. Birds blasting into windows. This is our world, and we are getting the same look in our own eyes. Our world is not the same world as in the movies, where no hero gets maimed or hurt, everybody finds love, and it is always somebody else who suffers the effect of the bad things. Horror films make us feel that bad things happen to other people but never to our safe and protected lives. There is a lot of talk about fake news, but there should be more talk about the fake world.

Watching the media and the politicians, you might get the impression that our (your) safety and health are their highest priority. That is what you always hear whenever anyone talks about it, but actions speak louder than words, and I, after so many years of experience, can tell you that your safety and health are not always the highest priority—and maybe not a high priority at all. Somewhere deep inside, we know that bad things do happen in our world. Each of us know somebody who had an injury or a death in the family, and it was not always from old age. It's too bad that we don't live in a world where we are protected, like actors or professional athletes do. The actual, real world, where real people are trying to survive or make a living, is a dangerous place, as you will see from my personal experience. Some people seem to be well insulated from some of the dangers, but even they can have unexpected injuries or illnesses. I think of celebrities who seem to have everything but really have more physical and mental problems than most of us. Then, as we go down the food chain, we find more exposure to danger at each level of society.

The so-called blue-collar workers—factory workers, construction workers, railroad workers, and tradespeople—work closest to dangerous machines, chemicals, explosions, fires, and a multitude of other fearful things. Consider all the dangers in congested and deteriorating cities, factories, coal mines, and other dangerous environments in the trades. We, as humans, need tools to survive against hazards we know exist and that have caused damage to other humans in the past. Are we confident that somebody is looking after our well-being? A meteor might fall out of the sky, or we could be struck by lightning, but what about the things in the world that can be controlled and the accidents and injuries that can be prevented? In 1970, under the Nixon administration, the government became concerned about the losses of lives and the cost of injuries and illness, and they created OSHA, the Occupational Safety and Health Administration, as well as NIOSH, the National Institute of Occupational Safety and Health, to make workers safe from well-known hazards that were causing death and serious harm. They also created other regulations, such as the Clean Air Act and Clean Water Act, to try to protect us and our families from harm to the extent that it was reasonable and possible.

At the time, it seemed like an easy fix and a great idea. Who knew that fifty years later, we would find ourselves with more anxiety than ever about the safety and security of the earth upon which we live? Like many good ideas with good intentions, sometimes the best intentions don't end with real progress, and things don't turn out the way we hoped. Things are better than before—there is no doubt. We don't have the black lung we did before, and we have contained asbestos. There are not too many big explosions, but the cost of injuries and illnesses continues to rise because of the cost of medical care and losses to families and society from unnecessary injuries and illnesses.

A look at the life of the safetyman will not change human behavior, nor will it make more resources available for the safety and health profession. What it might do is get more people to think about the world of danger and the culture in which we live. There are big problems we face in our modern society and questions about priorities. Right now, there are issues of immigration and the conditions where undocumented people are placed when they are captured in America without the required legal documents. We have an increasing disparity between the wealthy and the poor, and it looks like our infrastructure is not getting much upgrade or improvement. There are real concerns about the changing weather and the effects it has on the human population. In view of all these problems, the condition of a stairway, a guardrail, a bridge, or an electrical system might not always be the highest priority. A safetyman can't change those priorities or the world, but he can try to get the attention of individual people and families and make them aware of the problems they might face in their lives on a regular basis.

WHY DON'T WE CARE?

IT IS NOT EASY FOR ME TO SAY THAT WE DON'T CARE ABOUT other people. One of my friends and associates says, "Until it's them or theirs, nobody cares." That is just about right. We like to think of ourselves as gentle, caring people, but is that really who we are? How many genuinely caring people do you know? The truth of the matter gets back to our animal nature and our roots for survival of the fittest. The history of humanity and of our planet Earth was that when it was a big planet with few people, our selfish nature didn't matter too much. It didn't affect the planet, and it didn't affect other people. Advancement of the population and the development of technologies changed the picture. The small, self-serving actions of a few were not a big problem until there were so many people and so many chemicals and so many hazards. Our planet got smaller and smaller, and the hazards got bigger and bigger, and now we are in real trouble unless we change our ways or find another planet. Now we have climate change, plastic filling the ocean, and chemicals in our dwindling water supply. We can't seem to change our ways, and many people think we are on our way to oblivion. Our personal convenience and comfort are more important than the future of our planet, and the human suffering we see is not enough to get us to collectively act for our own protection. Those of us in a position to see the bigger picture and ring the alarm are not enough to get the masses to act when they are struggling to survive. The question of what we can do to try to change human nature to make people more caring is a difficult one. I personally do not think that people will become more caring or loving on their own. To me, it is the job of society and government to mandate behavior that protects the entire globe.

ARE THERE ACCIDENTS?

I THINK OF THE WORD *ACCIDENT* AS A MISREPRESENTATION OF the safety and health profession. It starts us at exactly the wrong place. The idea is that things will happen, and it is nobody's fault. My experience tells me the opposite. It's almost always somebody's fault. I believe those car and truck "accidents" can be avoided. If we spent as much time trying to find the causes for those accidents as we do adjusting and litigating them, we would stop having them. In my career, I have seen virtually zero accidents and injuries that I believe could not have been avoided. The industry and the financial resources in the area of accidents and injuries are just too large to eliminate now. Some people might ask about natural disasters or acts of God, and I must admit I do not think we can eliminate hurricanes, tornadoes, or earthquakes. But there are things we can do to mitigate the damage and protect our people better than we currently are. A tree might suddenly fall and injure or kill a family member. I think about why we are not inspecting the trees or why we allow camping or parking directly under an old tree.

I believe the word *accident* steers us in the wrong direction. The definition of an accident is "an unfortunate incident that happens unexpectedly and unintentionally, typically resulting in damage or injury." I have a problem with this definition and have learned not to use it even though it is in the name of my company. I am not going to be able to avoid using the word in this book because it is so widely used in the industry. The problem with the word *accident* is any implication that the incident couldn't be avoided. Words can be powerful, and anyone who thinks the word *accident* implies that the incident could not have been avoided is wrong. I choose to use the word *incident* instead of accident to avoid any misunderstanding. This does not mean that I think every incident of an accident can be avoided. There are things beyond our control. I do believe that 99 percent of accidents and incidents can be avoided.

WHAT IS SAFETY
AND HEALTH?

IN A FEW WORDS, IT IS YOUR LIFE AND QUALITY OF LIFE. IN the past, primitive humans hunted for food in a hostile world where all the species were fighting for survival and domination. In modern times, we have dominated all the other species, domesticated animals, and farmed for our food. In the process, we began to develop rules, starting with the Ten Commandments and the Code of Hammurabi from Mesopotamia, dating to about 1754 BC. This code was carved onto a massive finger-shaped black stone that, among other things, established punishments for violations of the code. Now, of course, it's the Occupational Safety and Health Administration (OSHA), American National Standards Institute (ANSI), and American Society for Testing and Materials (ASTM). In addition, and maybe more importantly, are the standards, customs, and practices within our society, which are what we understand to be the proper and correct way to do business and operate government (also known as mores).

They are what the courts decide when there are disputes regarding issues and differences of opinions.

In our modern society, we have come to expect cooperation between humans to not intentionally harm one another or create situations, materials, or equipment known to cause harm. The OSHA Act states that the "workplace will be free from recognized hazards." As a safetyman, I expand that definition to say the *world* will be free from recognized hazards. In OSHA, it says that the work environment must be inspected by "competent persons" to include the jobsite, materials, and equipment. As a safetyman, I expand that definition to say the *world* must be inspected by competent persons. In the end, we expect that the world should be free of hazards and that no person or company will intentionally allow hazards to exist or fail to correct hazards where they have knowledge of the hazards.

Many people believe that it is entirely up to the individual to look after themselves and their families. They take it for granted that it is a hostile and dangerous world, and precautions should be taken to the extent possible. If somebody intentionally injures a person, they should be prepared to defend themselves or suffer the consequences. Safety people believe that society has a responsibility to have order, and part of the order is being organized in a way that is protective, at least to the extent practicable. Most things in our world are organized in a protective manner. Traffic control is a big item not only for vehicles but also for everyday activities. Traffic today is more congested than ever, and the complexity of the interaction between different modes of transportation and pedestrians is substantial. I have seen horrible traffic control and significant life-threatening hazards on private property. We all expect traffic lights and signs to protect us from the traffic coming from different directions. We expect there to be safe roads, signs, signals, and barricades to allow us to travel safely.

We have speed limits that are largely ignored. Many basic concepts of traffic control are being ignored. Ignoring these basic safety rules and protections causes serious problems in our everyday lives. Danger when walking, recreating, or driving is a major symptom of a national and international problem. Think about speeding. It is so basic. The rules have been around for a long time. There are signs posted everywhere. Nobody is paying attention. We don't believe in speed limits or the safety they provide for our families. We are only interested in avoiding speeding tickets.

It is baffling to realize the truth of our lack of interest in our own safety and health. So many of the protections we think make good sense on the roads, in parking lots, and at business establishments are not viewed as important until something bad happens to someone we love. This lack of concern and investment in a program and protection for our families extends into many other aspects of our lives. We expect safe food, safe recreational activities, safe water, and safe air. We must have a system to make certain that our roads, food, water, and air are safe. A safety professional must investigate these situations and ask important questions. Take the speeding situation. How many people speed? Why do they speed? Does speeding result in injuries and deaths? You might think that taking a good look at these questions and then taking action would hugely benefit society. The problem is that society cannot be trusted when it comes to speeding unless it is enforced. Everybody slows down when a police car is present. There is technology to solve the speeding problem. As soon as the cameras document the speeding and other types of abnormal or unacceptable behavior on our roads and highways, it will stop. The nature of humanity is not always to look after our own best interest. This is the job of a safety professional.

Human behaviors are complex, and the best we can do to solve our safety and health problems is to take them on a case-by-case basis. Take an automobile accident (incident). It might be the fault or failure of the vehicle. It might be the failure of the owner for not performing the required maintenance. It could be the fault of improper signage, or it might be the fault of one or both drivers. It could even be the fault of construction activities on the roadside. There are many potential hazards on that roadway that can lead to the incident. These are all called proximate causes. We need a safety professional to look at all aspects to determine the root cause and prevent a reoccurrence. Who has the skills and knowledge to look beyond the fact that one vehicle ran into another vehicle? If the highway was built on a curve or the sun at a certain time of day blinds the driver, we need to know this and investigate it.

Determining both a reasonable level of safety and health and the cause of failure in instituting a measure of protection is complicated and requires evaluation, investigation, and the use of measures. This is the job of a safety professional, not just a police officer or an insurance adjuster. Too little effort and attention are focused on root causes and prevention. I have seen investigators not look at all proximate causes and root causes; therefore, there is never any progress toward a solution that would avoid further incidents. This doesn't make much sense and causes significant pain, suffering, and financial loss.

A safety professional is in a unique position to solve these problems for the benefit of every party. The police officer or the adjuster is not going to make recommendations for improvements for signage or road conditions. A safety professional can look at the history of this road and see if there have been similar incidents. They will do more than issue a citation and call the insurance company. A safety professional is aware of the potential causes of an incident, including engineering defects and human limitations. They have the investigation skills and knowledge to solve the problem—the root cause that led to the incident. A safety professional has the ability to utilize their experience, background, and the internet to find and investigate all the variables that might have caused the incident and then recommend specific action for resolution of the problem. They have access to standards, such as building codes, electrical codes, utility codes, fire codes, city codes, and municipal codes, and are aware of the customs and practices in the field. The safety professional pulls the problem up by the root so it will never grow again.

Before the internet existed, a safety professional of the past needed to have an extensive library and spend time finding a resolution. With an understanding of human behavior and the codes that represent the basis of standards, customs, and practices, a safety professional knows how and when standards apply. Speed limits don't work. What else can we try? After understanding the human behavior and determining and applying the proper standards, a safety professional needs to draw conclusions and form opinions regarding potential meaningful actions. They must apply tools of evaluation, such as logic, deductive reasoning, experimental methodology, and statistical analysis, to determine violations and render problem-solving opinions. Just because a safety professional believes something is true and has a feeling something is true, that is not enough. There must be some test to prove it. When a safety professional has developed opinions, the opinions must prove to be objective, without bias and not involving any kind of speculation. If a safety professional sticks to the facts, utilizes evaluation methodology, and has the credentials, experience, and communication skills, the work that they do makes a significant difference in preventing injuries and illnesses. This is more important than issuing a ticket or adjusting the damages. It is a whole different thing.

METHODOLOGY

METHODOLOGY IS LOOKING AT A PROBLEM AND TRYING TO FIND solutions that will work. It makes no sense to spend our time on speed limits when nobody believes in them. There is no reason to rush out and eliminate speed limits. The question is more about why they don't work and trying to find something to make them work—or at least make them work better. Cameras should identify speeders, and they can be ticketed. This does not seem to me to be a violation of civil liberties, but I am not a lawyer, just a safetyman. I will conduct an experiment to determine if my idea will work. It would not be difficult to compare the speed of traffic with and without visible cameras or the speed of traffic with or without tickets. We could look at actual incidents and see the significance of the decrease or increase based on the cameras and the tickets.

Methodology is the answer. Methodology is a system of methods used to form a theoretical analysis of the body methods and principles associated with the safety and health profession. The methodology is used to collect information and data for opinions. Interviews, surveys, and research techniques are commonly used in the process. The best-known methodology is called deductive reasoning. It is a cause-and-effect concept that involves collecting all the potential evidence and information and using logic to come to conclusions.

The next phase of methodology may involve testing and demonstrations. Once a conclusion or opinion has been reached, it may be tested to determine validity and reliability. Validity means that a test or demonstration shows that what was determined was correct and related to the actual incident. Reliability means that we can conduct the same test and demonstration repeatedly and will get the same results. There are many other methods used to analyze data, including fault-tree analysis, which allows diagrams to be used for all potential root causes and proximate causes. There are also what-if scenarios that are used to look at all possibilities beyond the ones that may seem obvious to the investigator. Finally, there are statistical analyses. Many statistical analyses employ the normal probability curve to compare events to see how likely they are to occur in relation to what is a normal probability. Many people call this the bell-shaped curve.

The important part of having a methodology or method is providing useful information to prevent a reoccurrence. Using random numbers, normal probability, and statistical analysis, we can find answers to some complex issues. For example, a question might be asked about how often a person might forget to turn off a piece of electrical equipment and the relationship of the failure of a human to turn off that equipment to the number of times a piece of electrical equipment, which is not turned off, causes an electrical fire. When I was in graduate school, we used computers and statistical methodology to determine the best location for emergency stop buttons on machinery. We wanted to know if it would be best to locate them up or down, left or right. The question was if we could determine the fastest method for an operator to reach the stop button. We simply connected their hands to the computer and tested the reaction time for the various locations. We found a correlation between distance and location and left- and right-handedness. The stop button should be as close as possible to the worker's hand and located on the side of the dominant hand.

I have always been interested in the relationship between maintenance and failure of equipment. There is an argument between "don't fix it if it isn't broken" and the whole concept of preventative maintenance. A study was done many years ago that showed it was more efficient to replace lightbulbs on a regularly scheduled basis than to replace them after they have burned out. I am not sure if there is any relevance to that study today with all the new lightbulb technology.

A HYPOTHETICAL EXPERIMENT

THERE IS A TELEVISION SHOW CALLED *WHAT WOULD YOU DO*. That show is a hypothetical experiment to show what people will do in a situation where they are able to do the right thing or the wrong thing. The only problem with the show is that sometimes the people who do the right thing know that the cameras are running. It could be that there are more people willing to do the right thing than we think there are, but I doubt it. We can find out by setting up experiments. For example, will people pay to have hazardous chemicals disposed of properly, or will they just throw them down the drain or in the garbage? I believe that most people will not pay to dispose of their hazardous chemicals or computers, and they will throw them away. I could be wrong. We could put tracking devices on the chemical containers and computers and find out, but then there may be a claim that privacy is being violated. I am always suspicious of people who are worried about their privacy being violated because I wonder what they are hiding. I can think of hundreds of privacy-violating experiments that could be assembled that would show the nature of human behavior as it applies to safety and health and avoidance of hazardous conditions, but every time I want to set up the cameras or the tracking devices to get the data, I need to prove that I am not violating any privacy laws, because people do not want their bad actions recorded.

As a safetyman for so many years, I have come to believe that humans will not always do the right thing when nobody is looking. Cameras seem to make people more caring about other people and the planet Earth. This privacy thing is a real problem for evaluating behavior. To understand human behavior, we have to study it under conditions where the behavior is real. We need to be watching to make sure people exhibit the safe and healthful behavior.

EXPOSURE, DURATION, FREQUENCY

THESE ARE ALL TERMS USED IN THE SAFETY AND HEALTH PRO-
fession to describe potential danger. They are words here, but they
may be used to calculate the seriousness of a hazard and to prioritize
the correction of hazards in reasonable order. Exposure is the first to
get attention. At OSHA, we used to say there could be a room full of
flying razor blades, but if nobody could get into that room, there is
no exposure to the hazard. I always thought in the back of my mind
that somebody must have a key. There are other standards that involve exposure, with which I have not been completely comfortable.
There is one about rotating hazards like fans or nip points and belts
and pulleys being seven feet high and, therefore, not needing to be
guarded. I always said to myself that some people are over seven feet
high, and then there are so many ladders. I don't even feel comfortable
when hazards are covered with a guard that can be removed without
an interlock to turn off the power. Someone could open a guard, not

knowing there was a danger, or they might try to lubricate or repair while the machine was operating. There is a principle in safety and health that says if there is no exposure, there is no hazard.

Exposure can also be measured in terms of the quantity of the material. This might be measured in the number of razor blades or asbestos particles. It could also be measured in parts per million (PPM) in terms of vapor or gases. Duration is secondary to exposure, and it is the amount of time the exposure exists. In the case of asbestos, everybody is getting a small amount of exposure, but if people are regularly exposed for longer periods, they should get more attention in terms of prevention and medical evaluation. Frequency is also secondary to exposure in that it is an indication of how often an exposure may occur. Is it once a year or once an hour? This concept has more applicability to health than to safety because we all understand that a single exposure to a safety hazard can have dire consequences. A single exposure to a health hazard, even a hearing hazard, can also have dire consequences, but we need to know the magnitude of the hazard, such as how loud the noise is, and the frequency of the hazard, as in how often it occurs. This means that calculating the seriousness of a danger or a hazard can give people the wrong impression. Noise is a great example. Noise is everywhere, and everybody is exposed to noise, but we need to know how load and how often to evaluate the danger.

In addition, the effect of noise on hearing loss is far from an exact science. There are individual differences, and there are different effects at various frequencies. There are also effects that vary from the noise that will blow out your eardrums to noise that will damage your hearing beyond the human perception. The safety professional must determine the best use of their time in solving hearing issues, and the guide for such a determination involves exposure, duration, and frequency.

Asbestos is similar, although it can be more significant from the standpoint of physiological effects. Some people might become seriously ill from a lower-level exposure than others who have exposures over many years. The best we can do is look at places where exposure, duration, and frequency are the greatest and work on reduction in those places. Some of those places might be where asbestos is naturally occurring. I once went to a place in Montana where there were high levels of asbestos coming from a mountainous area.

The bottom line is that these terms used to investigate and solve safety and health issues are not perfect, but from a scientific and methodological perspective, they are the best we can work with at this time. This is one of the reasons that research, such as is being conducted at the National Institute of Occupational Safety and Health (NIOSH), is so important. We might find out that certain types of noise are more important, or certain exposures to asbestos even at small exposure levels need more attention. There is a story in the news about Americans in Cuba being injured by exposure to noise that is allegedly doing serious damage to their brains. I have never heard of such noise, and I would have to ask NIOSH if this is even possible.

SAFETY EDUCATION

SAFETY AND HEALTH ENCOMPASS MANY FIELDS OF EDUCA-
tion—physiology, scientific methods, statistics, anthropometry (the
study of human body capacity) physics, ergonomics, chemistry, and
engineering, just to name a few. There are safety inspectors, and
there are safety experts. Some safety people do training on the standards, while others do investigation and research. It is a broad field
that continues to evolve. The evolution has created areas of study
and interest that are new and exciting. Twenty years ago, there was
very little study of ergonomics, and now it is the centerpiece of an
entire industry. As the study of safety and health has evolved, it has
been difficult to determine the best methods of education or a standard career path. Unlike the legal or medical professions, the safety
and health field did not have established schooling and testing for
the purpose of determining competency. Organizations, including
universities and community colleges, have been scrambling to set up
curriculums and certification programs. The equivalent of the bar
exam for being a safety professional is passing an examination to

become a certified safety professional (CSP) or a certified industrial hygienist (CIH).

The qualifications to become a CSP are as follows: a minimum of a bachelor's degree in any field or an associate's degree in safety, health, the environment, or a closely related field. The associate's degree must include at least four courses with at least twelve semester hours or eighteen quarter hours of study in the safety or health field. Becoming a safety professional involves education and experience that lead to the testing. The testing is then graded on a curve and followed by continuing-education requirements and recertification every five years. I personally would want my safety professional or industrial hygienist to be a certified safety professional (CSP) or a certified industrial hygienist (CIH). You will find that many, and perhaps most, safety and health practitioners are not certified. Following the certification process, there can also be a specialization process that may also involve certification, but many of those credentials and certifications are not fully established.

People want to know how they can become a safetyman like me. There really is not a good answer. It takes both education and experience, but I would say the thing that sets me apart from many other practitioners is my experience with experimental methodology at the University of Wisconsin. It was at the university where I learned how to examine variables and test hypotheses, which has helped me in all aspects of my work. On the other hand, all that study didn't give me the experience I needed to solve actual problems. If you have never worked in a steel mill or a foundry, it is difficult to learn of all the potential dangers and hazards to which those workers are exposed. In my opinion, the experience of the safety and health professional is not evaluated enough. Experience is helpful for success and credibility. There are many good safety and health degree programs that can be found in universities and colleges, especially at community colleges. This is an advancement from the era when a few safety courses were offered in psychology. At the same time, there continue to be many aspects of safety and health that involve human psychology, and it remains an important component, but now most safety and health programs reside in engineering.

There is still no replacement for the workers who understand the safety principles and concepts and apply them to the actual activities that they experience in their own workplace. An outside safety professional must utilize these inside people if they expect to make significant progress. There is a need for a combination of field experiences and study in several areas, such as physiology, statistical methods, and human factors. There are too many different hazards, jobs, and machines to describe a solution that will solve every problem. Fundamental is an understanding of the human condition and how we can survive without injury or illness in a world of hazards. With all the academic training there is today, there needs to be more recognition and training in human factors.

I remember when I started in the field at OSHA, I was sent for several weeks to a major Midwestern university to study hazard recognition. This was an educational program based on the standards, with pictures to show why a standard exists. This was important because a graduate of a university or college might not have an understanding of the components of a grinding wheel and the importance of the tools rests, flanges, operating speed, tongue guard, and the side guard if they don't understand the possibility of the grinding wheel exploding while it is engaged. Initially, I was surprised to find out they don't teach safety engineering or safety psychology in traditional schools. It has been necessary to have additional education in safety topics to become proficient in the field, and most of these specialized classes are offered at the community colleges. The vague path to the safety and health profession has opened an opportunity for many other people who did not follow traditional academic career paths.

Practitioners of safety and health come from numerous backgrounds. They might be academicians or might come from the trades. Many come from corporate management, and there are plenty of practitioners who believe safety management is management. Some practitioners have just a few OSHA courses, and a few have a single OSHA card. To be a successful safety professional takes extensive training in all the areas mentioned above, and you need to be a good communicator.

Many practitioners started by conducting training courses on the OSHA subparts. Much can be gained from these experiences, but a truly educated safety professional will know their own limitations and when a specialist is needed. This field is too broad for one individual to know it all. It is much like the medical field, where you have specialization in various areas. One doctor is a dermatologist, and another is a heart specialist. There is still room for a generalist who refers patients to the specialist. This is how it should work in the safety and health profession. Over the many years of practicing in the field, I learned to turn to specialized expertise when needed. I believe that is one of the characteristics that helped me succeed. A structural defect requires the services of a structural engineer. A mechanical problem requires a mechanical engineer. As a safetyman, I can work with these people and assist them in the overall process of evaluation. As soon as I see a potential design defect, I need a testing laboratory and a specialist. Too many safety professionals try to do it all and end up working outside their expertise. Knowing when you don't have the answers and need help is one of the most important parts of being an educated safety professional. I have learned to appreciate professionals who know their limitations.

WHY SHOULD YOU CARE?

WE LIVE IN AN INCREASINGLY DANGEROUS WORLD, WITH HOS-
tility and anger growing daily. We have climate change, road rage,
terrorism, and less control over chemicals and products entering our
stream of commerce. Products come in from all over the world without
evaluation or control. To get through this new world frontier, we have a
plan for survival. This safetyman idea involves the need for each of us
to be knowledgeable and educated enough to scrutinize the products,
material, and environment to know what hazards to avoid. People are
angrier and less satisfied than they have been in the past, and people
who are angry and unsatisfied become more aggressive. The world
has moved on and is not the same place it was in the past. The safetyman has been to places where these changes are taking place and
knows that individuals must be more protective of their exposure to
dangerous situations and hazardous environments.

Safetyman's knowledge includes knowing to stay away from a
specific area in a city or town where it is more dangerous to be at night
or even during the day. When you go to a different city or town, you

might not be aware of areas that are not safe to visit, and you can get into trouble. Just like avoiding a dangerous neighborhood, we also need to avoid situations where we are exposed to safety and health hazards. Are you boating in a clean lake? Is the property adjacent to a factory or a dump? We must know about chemicals and physical hazards that could shorten our lives or reduce the quality of our lives. Today, wherever you go, you have the potential of hidden danger.

The idea for the safetyman book is to make people aware of the danger that exists in the world and workplace. This should put us in a better position to avoid dangers and allow us to have longer and more fulfilling lives. As a safetyman, I have found dangerous and hazardous situations, and the information we gather needs to be shared. Today, with the internet and social media, we are in a better position to share important information. You may not think that there will be poison in the air or in the water. You may not think that a crane or ice will fall from a building. You might think that people are looking after your safety when you go to a store or ride an elevator or escalator. You might think the traffic control will make your commute safe for your family. You could be wrong. The safetyman gives you information and tools to survive in an increasingly hostile world.

When you shop for consumer products and compare safety ratings for a child's car seat or a bed, or when you buy clothes that are nonflammable, you are being a safetyman or safety woman. When you buy quality tools, not the least expensive ones, and read warnings, you are being a safety person. It is one thing to go into the world and assume the world cares about you and your family. It's another thing to be proactive and assume nothing. There will be dangerous people and situations where safety has not been considered at all. You should be suspicious of the products, material, and people you think are looking after you and your family.

BATTLING FOR SAFETY

I THOUGHT EVERYONE WOULD BE ON BOARD WITH SAFETY AND health, but I was wrong. I thought people would obey speed limits. I never thought that people or institutions would dump hazardous materials in our oceans, rivers, and streams. Many people are against all regulations, safety, health, and environmental rules. They believe in taking chances. People think that establishing and following rules impinges on individual liberty. This seems wrong to me because we are so often victims of unseen and unknown danger in our world. We can become victims of our own human flaws. It is still surprising to me when I talk to people who think safety is an enemy or burden. Sometimes I think that safety is less of a profession and more a battle between good and evil.

Yesterday, I got the same question I get repeatedly. The question is about the mommy state and how we need to take care of ourselves and not have the government look after us. I immediately think about a world without traffic lights, and I want to shake the questioner and ask them what they are thinking. Everywhere we go, we are protected by government regulations and safety and health professionals. Without them, we would all be in trouble.

The next question is about how far we should go to protect people. Do we need to watch everybody? It reminds me of an incident that happened in a hospital where there was a puddle of water on the floor. A woman slipped and had to have her leg amputated because the compound fracture cut her artery. The hospital said they couldn't watch for water on the floor every minute. I looked for an engineered solution. I suggested a mat or carpet could be placed under the drinking fountain where people were most likely filling up water bottles and the water was dripping on the floor. I was trying to explain to the hospital that a competent safety person would see a source of water used by the public next to a linoleum floor and understand the problem. This must be done before somebody slips and gets hurt. The hospital knows they are inviting old, sick, and infirmed people, as well as people on crutches, and they should know they need professionals to look for potential hazards before an injury happens. This is what it means to be proactive for safety. I know when I talk to the hospital that I can make them understand, but I still have to have this fight with people every day who are saying (1) they don't want to be proactive for safety, (2) it's too much trouble to do safety, (3) or it's too much money to do safety. It's frustrating because, in the end, it is much cheaper to have safety than to pay for an amputated leg. The fight for safety is a big deal in my life, and it never ends. The safetymen and safety women of the future need to fight for safety and health every day.

SAFETYMAN AS A SUPERHERO

I THINK BEING A SAFETYMAN OR A SAFETY WOMAN IS BEING A superhero! We are fighters! You may not realize how hard it is to do this important work. We look for trouble before it happens and are obligated to get things changed before something terrible happens. Every day, everywhere I go, I see hazards and violations. From the time I leave the house until I come home, it is a never-ending inspection of missing fire extinguishers, poor wiring, tripping and slipping hazards, people using ladders and scaffolding the wrong way, a trench without a trench box, people hammering nails without safety glasses, and more. If I stopped for every hazard, I would never get anywhere.

Yesterday at the big-box store, I saw pallets of bottled water piled at least five levels high, thirty feet in the air. One of the pallets at the top was leaning toward the aisle. It could fall, and if it did fall, it would kill a human being. The people with me were annoyed when I felt the need to contact management and get the tower of bottled water lowered to a safe level. Of course, the question is, why in the world did the forklift operator stack those pallets so high and leaning into the aisle? If even one of those water bottles on the aisle side were to burst, the entire pallet would fall. I could see there was not enough space in the warehouse to store all that water. The root cause of this dangerous hazard is a storage problem. I know that it is the sales floor where the money is made, and the storage areas are just overhead. This is not only a storage problem and an inventory problem but also a training problem and a business problem. The big-box store needs to figure out how to keep less inventory or how to get more storage area where the inventory can be safely stored. There is a specific standard in OSHA and ANSI for the proper storage of materials.

I was recently at the commuter train station. I was investigating a fatality where a welding rig blew up in a train truck. The door blew open, and the explosion was so severe it killed one worker and injured another employee in the area. The welding rig was being transported with the leads (the connecting hoses) and the pressure gauges attached to the oxygen and acetylene cylinders. This is against regulations because it means that these tanks, which are under pressure, will leak and create an explosive mixture that could collect in any enclosed space. A third worker was grinding and creating sparks in the same area, and when those sparks contacted the explosive gas, there was a terrific explosion. I went to this site to conduct an inspection with the expresses purpose of making recommendations to prevent another explosion. I could see how they got into trouble by not taking the time to remove the hoses and gauges each time they were transported on the truck. I emphasized the importance of removing the hoses and gauges and replacing the valve protection caps, and the people at the railroad started getting angry at me. As I have found in other cases, it was clear that these people dislike the safetyman.

I am a friendly, gregarious person who is trying to save people from death, pain, and suffering, and not only don't they appreciate my work, they act like they wish it was me who got blown up in the explosion. In my regular life, I try hard to get along with everybody. As I was investigating this fatality, I was in a room with flammable storage cabinets. I opened the door to these cabinets, and they were filled with flammable liquids. One of the cabinets was stacked high with combustible boxes, and another was completely blocking a major electrical panel. These are all violations. There was also a pit for working under the trucks within five feet of these flammable vapors. This was an area where they already had one explosion, and now I was in a situation where solvent vapors, heavier than air, were leaking into the pit, where any spark or even a static spark could send us all to heaven. I told the top manager about these violations and about my concerns, and he became even more angry. I was giving him a free inspection, trying to protect his plant and employees and myself. I normally charge for these inspections. You might think that these people would have at least some small appreciation for my work or my experience. There was no appreciation at all. This guy said to me, "Please try and stay in your lane," which meant to focus on the investigation of the fatality, not on other dangers that existed at the facility.

I told him I had an obligation to protect people and property, and he said, "If you stay on the ice, I will have to put you in the penalty box." It felt like a threat, but I let it go. I felt the need to report the situation to higher authority before there was another incident. I talked to the top manager in the department, but there was no indication that they had any intention of correcting the violations.

I am reminded of another investigation I conducted a few years ago at that same commuter railroad. They were going to enter a confined space with zero preparation, and I shut down the job and told them if they entered that pit without testing for oxygen, I would call the police. They were so angry that I was slowing them down and telling them how to do their job that I honestly thought one of the managers was going to hit me. I am rarely treated with respect as a safetyman, and in forty-six years, I can only remember a few times when somebody said thank you or appreciated my efforts. Being a safety professional requires a person to be a real hero and as tough as nails. You must be fearless and stand up to powerful people every day. You must hold your ground despite people who are mean and may want to hurt you. I have had many experiences where I thought somebody wanted to hurt me. There was a time at OSHA (Occupational Safety and Health Administration) when I was at Wisconsin Steel, which is now closed because they had so many violations. While I was inspecting, they opened a snort valve on a blast furnace and perforated my eardrums. I thought they did it on purpose, but I couldn't prove it.

Recently, I went to a steel mill, and they sent workers with respirators to chip away at refractory material under a blast furnace that contained high levels of silica and asbestos. I was investigating an injury with no respirator at all. I ran out from under that furnace, and I was angry and shouting at the people in charge. They were smirking. And there was nothing I could do about it. It will take many years for that silica and asbestos to affect my body, and I will not be able to prove that these guys were trying to hurt me, but I can tell you that I was coughing for three days after the incident.

I am not bragging about being a hero. I'm just trying to make you understand how hard it is to be a safetyman. It has been extraordinarily rewarding both professionally and personally, but there are only a few safety professionals in the world. I have been told many times that after I retire, it will be difficult to find another safetyman like me. They ask me why I am not training somebody to take over my work. I tell them I know that safetymen like me are a dying breed. I feel like the last dinosaur, because most of the older OSHA guys like me are gone. We were a rough group, not afraid to stand up to anyone for safety. I don't know anyone to take on my job. It's too hard for most people to deal with all these aggressive people who do not cooperate with safety standards, and I don't know anyone with the background, personality, and stamina to do the job the way I would want to have it done. It worries me that there may not be safetymen and women to stand up to authority in the future. I have worked with many people over the years, but it is always a problem for me to find people with the knowledge and toughness required to get this job done. Most of the candidates get tired of everybody picking on them and hating them. Besides, there is so much more money going with management.

When I first started the safetyman concept at OSHA, many people didn't like it, or they made fun of it. Now everybody calls me the safetyman. It started with my now-departed best friend, Bill. While I was working at OSHA and in safety, Bill always called me the Wise Old Owl of safety. Every time he said it to me, I thought that, in fact, I was a safetyman. From that time on, I had a unique identity. One of the best things I ever did was to protect Safetyman.com when the internet was first started. That turned out to be a good move because as I've gone along in my career, people have always been able to find me. I think of the web page as being like the bat signal that the commissioner would send out for Batman. Sometimes people think of me calling myself the safetyman as egotistical, but I don't think that's fair; it's just that they don't know me. If you investigate me, you will see it is just the truth.

I see my safetyman life as an adventure because being the safetyman means acting differently than all the other people. I wasn't bitten by a radioactive safety spider, but I did have some adversity that led me to be the one true safetyman. At the time of this writing, I have been the safetyman for most of my life. I have a safetyman website and email. I even have a "Safetyman Song," which some people think is silly. The "Safetyman Song" wasn't meant to be silly at all. It was an attempt to recruit others to be safetymen and safety women so we could make the world a better place. I have tried as hard as I can to make the world safer and more healthful, but it is not easy to fight the forces of nature and society that tell everybody to "just do it" and to do whatever you want or what feels good. I learned along the way that being a safetyman and getting others to be safety people is a difficult task. Working together, we could save an enormous amount of pain, suffering, and money. Even if you are not sensitive to all the pain and suffering, saving all this money should be enough to bring attention to the value of the safety and health professional.

The National Safety Council used to use a picture of an iceberg of the hidden cost of accidents (I still prefer to call them incidents, injuries, or fatalities because accident implies they can't be prevented). At the tip of the iceberg, the part that shows over the water, there was the actual paid costs of an injury, but underneath the water were the costs of replacement of the worker, downtime in production, training costs, OSHA costs, litigation costs, and insurance costs. That bottom portion of the iceberg that consists of medical treatment and litigation has skyrocketed. We all know how much these costs have escalated because we pay the costs with our insurance premiums. In addition to insurance costs, the medical, legal, and expert costs have risen exponentially over the years. It is a great joy to save people from injuries and illnesses, but there is also joy in saving all that potentially wasted money. Being a safetyman is more than a job or a career. When you call yourself a safetyman, you are a safetyman twenty-four hours a day every day of the year. I never set out to be telling others about the value of safety and health or to be called the safetyman.

I remember wanting to be a safetyman even before going to work for OSHA. I was trying to call companies to get them to understand the value of having a safetyman and telling them how I could save them money on workers' compensation, injuries, and illnesses. Back in 1974, it wasn't working, and I needed to drive a taxi just to pay my bills. It was frustrating, but I wanted to devote my life to something worthwhile, and being a safetyman seemed like a good cause. I wanted to be able to help people who needed help and make a living at the same time. I didn't know what else to do to continue my effort, so I decided to take a job at OSHA. At first, it seemed like the wrong path, as it wasn't my intention to be a government worker. Because I had received OSHA training from the state of Wisconsin while doing my graduate work, federal OSHA was interested in hiring me. It turned out to be the best move I ever made. To this day, I tell aspiring safetymen and safety women to work for OSHA because it is the only place that will give you the diversity of training and experience you need to be successful in the field.

Once I got to OSHA, every opportunity was given to me to get training and to have the experience to be a professional safety engineer. I loved it so much that I intended to stay at OSHA for my entire career. What I never understood until I left OSHA was the independence that working for a government agency can give an aspiring professional. Upon leaving OSHA, I found that I had a neutral perspective on the business of avoiding accidents and injuries. I wasn't ever in a position where I had to take sides on safety issues. OSHA developed me into an independent evaluator of hazards.

Once I left OSHA, I had several different opportunities because safety and health had become more important and prominent, and the costs of injuries to insurance companies and private companies were rising rapidly. I realized that when a single injury is avoided, it can mean tens of millions of dollars are saved, and when the injury happens, there can be millions of dollars in costs, including litigation. It is important to understand that when a young man or woman is seriously hurt, it affects a family so significantly that there is no alternative to litigation. That family still needs to eat and have a place to live. When a worker suddenly stops working, in many cases, through no fault of their own it's either litigation or welfare. I have seen people who were injured on the job, because of poor equipment or safety practices, refuse to go to a lawyer. They lose their spirit and become depressed and overwhelmed with self-pity. People don't understand that when they get older, they will need more treatment. As time goes by, the costs will rise, and they will need the resources to survive.

Whether it is trying to keep these terrible things from happening or working to get a family protected from a life-changing event, a safetyman is involved and valuable. I am astonished that more young people are not trying to be safetymen and safety women, as it is a great and rewarding career. The difficulty in becoming a safety professional is far surpassed by the rewards of making a nice living and the feeling of accomplishment you have from knowing you are trying to improve the human condition. One aspect that takes getting used to is the horrible injuries and the problem of being disliked for doing your job. People do not like to be told what to do, and they especially don't like it if you are pointing out that they are doing something unsafe. The worst is when you must point out that they are doing something dangerous. Many people still believe that common sense will take care of everything, and danger is open and obvious.

Being a safetyman is a unique occupation—so unique that National Geographic television once called me about a show, for me to narrate about dangerous occupations. I told them that the friction and drama between a safetyman and the rest of society would be the star of the show. The show must have fallen through, but it could have been a hit. We would be going to different industries all over the country and the world, showing the battle between the forces of production and profits versus safety and health. We would see the danger people face and listen to what they have to say about it. I would tell the audience the background, statistics, and standards involved in the dangerous work. Ironworkers love the danger of climbing on steel, hundreds of feet above the ground, where one false move is the end of their life. Every ironworker knows someone who has lost their life, but they still do it and like the danger.

To this day, OSHA has an old standard for steel erection that allows ironworkers to work without fall protection fifteen or even thirty feet above the ground. I like to say those ironworkers must be tougher than the other construction workers to be allowed to fall from that height. To me, it's discrimination. I have argued with ironworkers about this standard because it's just plain wrong to subject humans to falling fifteen feet when there is perfectly good hardware that would protect them from the fall. Years ago, they didn't have the stanchions that attach to the steel and act as guardrails and which the ironworkers can tie off and be protected by a harness and shock-absorbing lanyards. The problem is that these advancements in technology don't matter. This group wants to have danger in their lives. It's no different from a motorcycle traveling at ninety miles per hour on the highway and the driver not wearing a helmet. When the motorcycle driver crashes, he and his family face the consequences. When an ironworker falls to the ground, we all end up paying for it. No matter how macho these guys are, once they fall, they have lawyers suing everybody to get money for their family. It's a wasteful system that needs to be changed, but it's even crazier to hear that today young people will risk their lives to go viral on the internet. I see pictures of young people jumping out of automobiles or off bridges and having it photographed for their fifteen minutes of fame.

THE WORLD REAL
AND IMAGINED

YOU ASSUME THEY DESIGNED THE ESCALATOR SO THAT IT won't hurt children. A safetyman finds out there are openings on the side that have repeatedly grabbed small children and have done serious damage. There are videos on the internet of children lifted and dropped from the escalator railing. A youngster got his arm torn off in an escalator, and there have been fingers and feet caught in the moving parts. There is nobody supervising the use of the escalator by children or physically challenged people. Recently, twenty-four people were injured after a crowded escalator at a Rome metro station malfunctioned and suddenly sped up and hurled people down the moving stairway. Family members died after tumbling down a metro station escalator in Washington. This family thought it would be safe to carry a motorized scooter on the escalator. I have seen heavy luggage falling down the escalator and hitting people, knocking them down. We imagine a world with escalators that are safe, and we relish the idea of not having to climb the stairs with our luggage. I have not seen any good statistics about escalator injuries, but I am certain they are massive. So, the next time you are taking an escalator, especially if you have children or luggage, I hope you will think about your safety and the safety of your family. This is just one example of people believing that the world they live in is safe, when the real world is not safe at all.

Once again, it's like the speed limits on highways that are not enforced by the police. There was a time when we thought it was fine to go through the intersection on the yellow light. Today, things are a little different, with drivers increasingly distracted by their cell phones. Today, with increased awareness of distraction, we might think it's better to wait and proceed slowly. Changes in the world are making it more dangerous in ways that we never anticipated. It did not happen all at once; things have been getting more and more hazardous and dangerous for years, with more people, more travel, and more cars. People are living longer, working longer, and driving longer. The same technology that has eliminated some hazards has created other hazards, such as distraction. Even twenty years ago, nobody would think that people would be walking around, not looking where they are going because they are so distracted by their cell phone. Who knew that all these new drugs would be developed with all their side effects? Who knew marijuana would be legalized? In an increasingly hostile and dangerous world, it is more important than ever to have a plan and a program to survive.

OPINIONS

THE SAFETYMAN HAS LIVED HIS LIFE WITH THE ASSUMPTION that the avoidance of accidents and injuries will prevent the terror and heartache sustained by individuals and families and will save the enormous costs of losses. I believed all along that everybody would be on board for the prevention and especially for the savings. It turned out I was wrong. The problem is that people have different opinions about safety, and the fact that simple actions can be taken to prevent pain, suffering, and economic losses is not enough to change some people's opinions. Most people still believe that losses are due to the actions of individual unsafe people. The idea of survival of the fittest in the world or the workplace seems harsh to me, even though I fully appreciate the human characteristics that lead to accidents and injuries. There is a belief that these unsafe individuals should be held accountable by suffering the pain, loss, and economic instability that accompany accidents and injuries. We are not a very careful species and are subject and prone to a variety of frailties. People need to be safe despite these human limitations.

I recently came back from Houston, Texas, where I visited a facility where they were using ancient turret lathes to make metal parts. A worker had his arm torn off when he reached into a moving machine to retrieve a gauge used in the process. I was surprised to hear from everyone at the facility that the worker was at fault for sticking his arm in the rotating machine. There was general agreement that this worker was trying to take a shortcut to increase production. I found myself a little exasperated when I pointed out the safety standards that require rotating machines to be guarded or protected to make it impossible to get any part of the body in the bite or the danger zone. I suggested a fully interlocked guard, or two-hand controls would prevent this incident from happening again. The people looked at me like I was slapping them in the face. If I could read their minds, I could hear them saying that these items would slow down production. If they could read my mind, they would be hearing, "But what about this man's missing arm?" I wondered why a worker would be interested in taking a shortcut to increase production if there was no incentive for him to do so.

An incentive program used in many factories is called "piece rate." Workers are paid a fixed piece rate for each unit produced or action performed, regardless of the time it takes. Under both state and federal law, employers must pay at least minimum wage to employees, but one option has always been to pay a piece rate, as opposed to paying an hourly rate. The piece-rate rules are regulated by the Department of Labor to assure at least the minimum wage, and there are also provisions for breaks, and they must adhere to safety standards. The problem with the piece rate from a safetyman point of view is when the management does not adhere to safety rules and supervision is so poor that shortcuts are possible. The piece rate in this case is being used to encourage the worker to work faster to get more pay. If there was no danger of contacting the rotating parts, and the machine was fully guarded or protected, there would be no reason for a piece rate or incentive rate to be used at all. The machine would determine the number of pieces rather than how fast the operator could interact with the moving machine. The reader can understand that automation, computers, and robots take the piece rate out of the picture. There is no longer any incentive for a worker to take a shortcut or put their body in danger. In fact, it would be impossible to put their body in danger.

The idea of compliance with safety standards and the elimination of the motivation of workers to take shortcuts or chances seems fundamental to me. What I found in Houston was that most of the people I was dealing with had a different opinion. They don't understand or want to understand the requirements for machine guarding, and they think of these workers as competing against one another and, to some extent, against the machine for higher wages. That is until the arm is missing. The question in my mind is, *How do I get this company to spend a million dollars to upgrade the guarding on these machines (thirteen of them)?* It would be cheaper for them to just pay the fine and the cost of possible litigation, then hope and pray that another worker will not make the same mistake. After seeing what happened to this one worker, I have no doubt that the others will be more careful.

People having a difference of opinion creeps into my work and many aspects of our society. Some people don't believe in any regulation at all. Others are accused of being a part of a nanny state where they expect everything to be taken care of by the government. It's fine to have all kinds of opinions and differences in opinions until something bad happens to you or your loved one. When a situation is painful to individuals and families, and there is a significant financial incentive to intervene, it is my belief that we can all agree that we need to do something to prevent the pain and suffering and save all that money.

SAFETY HAZARD ANALYSIS

Everybody needs to do a safety hazard analysis. It's not enough that they do it at work. We need to do it at home to protect our people from falls, chemicals, and a multitude of other hazards. Statistics show most people get injured in their homes. We know the bathtub and the roof are places to examine. If you have children or pets, you need to lock up drugs and household chemicals. I have found that people in some homes don't lock up the power tools or even the guns. Most companies and organizations today have requirements for task analyses. This is the idea of identifying potential problems and hazards in advance and planning for solutions. We identify potential dangers to develop a plan to deal with a potential hazard. It's not complicated, and we can do it at home. A simple blank piece of paper will work fine: (1) this is what we are doing, (2) here is what we need to watch out for, and (3) here is what we can do to eliminate or reduce the danger.

MOVING THE COUCH TO THE BASEMENT

Moving the couch may seem like a simple task, but when analyzed, it can be complex and dangerous. Most will just get some help, lift the couch, and try to get it through the doorway and down the stairs. The problem is lifting the couch if you are not strong enough—the distribution of the weight of the couch and the logistics of getting it down the stairs under control.

TASKS

Lift the couch safely.

Get the couch down the stairs under control.

Place the couch in position downstairs without obstacles.

POTENTIAL HAZARDS

- weight of the couch—back injury with tandem lift
- size of couch versus the stairwell—won't fit or the body getting caught between opening and couch
- handling and controlling the couch—use equipment like a dolly
- placing the couch—easy access to the location

SOLUTIONS

1. Determine the weight of the couch. Will it come apart? (Take off legs and cushions.)
2. Measure the couch. Will it fit and in which direction?
3. Measure the doorway and stairway. Is the handrail in the way? (Can it be removed?)
4. Use a dolly. Lifting even with assistance is dangerous when changing elevation, as all the weight will be on the downside.
5. Tip the couch on the dolly. This will prevent the lifting.
6. Secure the couch on the dolly. Use straps
7. Arrange the location and placement of the couch. Make sure the route is clear.

Hire professional movers if you don't have a dolly, if you determine it won't fit through the doorway, or if there is any possibility of losing control.

A simple task is not as simple as it seems. As a safetyman, I always choose to hire a professional with skills, training, and equipment. It is better for my health and eliminates the possibility of a serious injury. Having a moving party and offering pizza to your family and friends could be a costly mistake. Don't invite me to your moving party because I will not show up. A problem happens with churches, schools, and organizations that want members to volunteer for tasks they are not qualified to do safely and professionally. Countless times, I have been asked to move furniture, paint on a ladder, and move hundreds of chairs from one location to another. You don't have to listen to me, the safetyman, but these tasks are dangerous and full of opportunity to ruin the quality of your life. It's your right to choose. I suggest you choose wisely. If you are not physically fit and if you are not lifting at the gym on a regular basis, please don't walk but run away from these tasks. It is always surprising to me how these church and civic leaders allow out-of-shape members to do work that inevitably leads to injury. One time, I was asked to help clean out a community center, and I went there with respirators to give to the people who would be exposed to the dust. What I found was not just regular dust but asbestos. I told everybody that they needed to leave and hire a state-certified and licensed operator to remove the asbestos. Nobody paid any attention to me.

Volunteerism is a big deal. I like the idea of loaning your skills to an organization. If you are a professional mover or painter, you can volunteer your skills. The same applies if you are a plumber or an electrician, but the idea of having managers or salespeople doing these tasks is against every principle of safety and health. One of the problems in these situations is that everybody wants to pitch in and feel like they contributed. Nobody wants to be the last pick or sitting on the sidelines while other people are doing the work. There is so much social pressure in these situations that when you think about it, an accident or injury is almost inevitable. I have seen it happen many times. Just picking up debris without the proper gloves or hammering nails without the proper safety glasses is dangerous, and even a wood sliver may become infected. Please watch out for the social pressure.

To combat these pressures, survival and well-being should be powerful forces. These survival forces don't seem to be strong enough to overcome social pressure. Overcoming social pressure is one of the biggest items for the entire safety and health concept. We must analyze situations in our daily lives that could result in life-changing injuries. We need a guide to make us think about what we are doing. With the information we currently have on the internet, we can go much further in protecting ourselves with a little planning and research. Just as we plan for the weather and for rush hour traffic, we can plan for moving object projects. We humans get into trouble because we are subject to social pressure, impatience, and impetuousness. One of the main ideas of this safetyman concept is to start with the actual characteristics of the human condition, and then we can have a basis for understanding problems that may be encountered and that could cause injury or illness.

DANGER AND HIDDEN DANGER

BEING A SAFETYMAN REQUIRES BEING AN INVESTIGATOR. ON all those crime shows on television, it's always a murder being investigated and never an accident or injury. The job of a safetyman is like that of the detectives you see on all those shows. It's an exciting job to try to find out how some unfortunate event was caused and to be involved in developing procedures to prevent it from happening again. In all those television shows, we are looking for the person or people who intentionally did something wrong. In the safety field, it is a broader investigation because we are looking into different kinds of danger. We get to investigate the difference between danger and hidden danger. Some danger can be seen clearly, like the cutting machine that is common in most industries. This machine has a sharp blade, and you should try to stay away from it. But there can be a different kind of danger that is hidden when you put a guard on the cutting machine or some kind of device that gives the worker the sensation that the machine may be approached safely. The unexpected happens, and the guard might not function the way it was intended, resulting in dire consequences. Having the illusion that something is safe when it is not safe is one of the biggest human problems we face. Car seats, cribs, and airbags are just a few examples, and prescription drugs are becoming a bigger issue.

Hidden danger is more difficult to find than regular danger. One of the biggest problems for the safety professional in engineering safety is being sure to avoid hidden danger. It is anticipated in the OSHA standards that a guard should never create a hazard. If a guard hides a cutting blade that can still be contacted or if the guard creates a nip or pinch point, adding a guard or safety device can create a hazard. Hidden danger by its nature needs to be discovered. An experienced safety professional gets a knack for seeing hidden danger. You might see a boat, but a safety professional sees sharp blades and the possibility of drowning. Will you be wearing head protection on the boat? Will there be drinking on the boat? Will there be waterskiing? Hidden danger looms over almost every activity. When we go hiking in Alaska, we always take bear spray, and when we go fishing, we hire the most experienced guides. Those guides have story after story of people from Michigan or Wisconsin coming to Alaska and not understanding the currents, the locations of the rocks, or even the waterfalls. I have seen boats run aground and bears attack runners and bicycle riders. You would think that tourists boating on the Kenai would go to the trouble of buying a map.

Most people are not looking for the hidden danger, and in some cases, they are looking away. I have had people tell me that my investigation into their (our) activities is ruining their fun. If we are going dirt biking, sailing, hiking, or camping, I have a lot of questions. The safetyman's wish is for every family to take some time to think about hidden danger. I believe that people should have escape plans and routes from their homes. Every home needs a fire extinguisher, a carbon monoxide detector, and an evacuation plan. When I see a bedroom located over a garage, I see a hidden danger. I can count on both hands the times I have seen people forget to turn off the engine in their car. One of my car's engine runs so quietly that I can't tell if it is on or off. With remote starters, it is even more concerning.

Hidden danger is present in so many areas of safety. Electricity cannot be seen, radiation is not seen, hazardous chemical vapors and gases are not seen, hot objects don't give off signals, and even rotating blades seem invisible. I have had two cases where a passenger—and in one of the cases, a pilot—walked directly into the tail rotor of a helicopter. A hidden danger can be a heavy branch falling from a tree. Do you sit under trees? When you go to the hockey game, do you know which seats have the puck flying in their direction at speeds faster than the human eye?

PSYCHOLOGICAL ISSUES

What are human weaknesses? What are human limitations? Not enough thought is given to these important issues. We know humans don't pay attention to warnings. Cigarette and alcohol warnings have not been a deterrent, while cancer has been more effective. Fear is a better deterrent than words. This is a simple but important lesson in safety. The problem is that we use words for most of our warnings—not even spoken words but written words that nobody reads. Thousands of pieces of elevated equipment have repeatedly been run into the overhead power lines. There are plenty of warnings on every piece of equipment. Operators' stations have a written warning, "DO NOT RUN INTO POWER LINES." This is not working! We need something else. A safety professional or safety engineer will point out the need for barricades or signal people to make certain that when the equipment is used, it will never run into the power lines. Good idea, but still, little effort has gone into the basic problem of people not responding to warnings. Wherever you go, you see a caution sign or orange cones telling you that the floor is slippery. These cones and signs do nothing to protect you from the danger. The world acts as if when you have a wet floor, all you need to do is put out a sign or a cone, and you are done. The result is many injuries.

My favorite is the one downtown where they warn you that ice is falling off the building, "Beware of falling ice," but they do nothing to protect you from the ice. I hope the reader can understand this basic problem and realize that this is not protective; we need something more. This thinking leads to the safety hierarchy that dictates that a hazard should be eliminated through engineering methods before consideration of any other effort. If a hazard is eliminated, it is gone. Signs, warnings, and protective equipment mean the hazard still exists. I am always surprised when a client uses a hazardous chemical and provides protective equipment. When I suggest that they use a different, less hazardous chemical, they often tell me that they never thought of that possibility.

SOCIAL PRESSURE

Social pressure is one of the most powerful forces in our human universe, and it has not been studied enough. As a safetyman and human factors expert, I see the negative effects of social pressure everywhere I go to work. People are very social and feel social pressures as individuals and in groups. Social pressure has been defined as direct connectivity on people by peers or the effect on an individual who gets encouraged to follow their peers' values and behaviors. Everyone knows about the pressure we feel to get married and have children, but studies have shown that social pressure can change a person's view of reality and make them act in an irrational manner.

Peer pressure is a part of the culture of every organization, and peer pressure may affect behaviors and the safety and health culture. Individuals encounter peer pressure to be risk-takers or to speed up production, impinging on safety and health efforts. The safety and health professional must examine the culture and attempt to influence the culture with concepts that put safety ahead of risk-taking and production. This is a difficult task because people can be set in their ways, and generally, they have no desire to be told what to do or how to behave. All of us have been victimized by social pressure. The phenomenon called the bandwagon effect is powerful: we don't want to be different from our peers. We want to fit in. If everybody in the group wants to ride a motorcycle without a helmet, it makes us feel like we want to be a part of the group and refrain from wearing the helmet. We want to be cool, and we want to be accepted. These attitudes can all lead to risk-taking behavior that is the bane of the safety and health effort.

I believe social pressure, peer pressure, and lack of safety culture are the main causes of accidents and injuries. Nobody wants to be viewed negatively by their peers. Nobody wants to be called names. It is not possible to speak about the problem of protecting lives without dealing with social pressure and safety and health culture.

There is a natural competition between individuals to be stronger, smarter, more attractive, and more popular. There are physical competitions for speed and strength that start in childhood. Many of us have had the feeling, in school and in adulthood, that we were not the first pick for the team or that we were weaker or not as smart as our peers. These pressures often weigh on our lives and can cause lapses in judgment that can end in catastrophic losses. I have interviewed thousands of injured people, and each one of them had no idea that something bad was going to happen when they did something dangerous or acted without thinking or planning. They always tell me that everybody else was doing the same unsafe act that they believed caused their injury. There can be a tendency to blame an individual for a single unsafe act, while the culture in society or at the facility has encouraged and supported the same risk-taking behavior for decades.

OVERCONFIDENCE

Carpenters lose fingers because they feel their skills are so developed that they no longer require a guard. People do routine tasks for years without a problem, and then suddenly, they try something different or something unexpected happens, and they suffer a devastating injury. People will reach inside a dangerous machine without turning it off at the source for no apparent reason.

Overconfidence is part of human nature. Nobody knows what it is like to face a devastating injury, and nobody wants to walk around this world fearful of every possible thing that could happen. The trouble with overconfidence is that there are so many things that are out of our control. One time, I was in Alaska, and we were watching small bush airplanes landing at a restaurant. I was with a good friend, and we decided to sit on a wooden fence while we watched the planes. Suddenly, the fence just fell over from our weight, and my friend hurt himself enough to visit the hospital.

Last Thanksgiving, I went to close the vents on my barbeque just as I have hundreds of times in the past, only this time I had to reach between two barbeques, as I had cooked two turkeys. I didn't have the same amount of room to access the handle to close the vents, and I brushed my forearm across the kettle and got a nasty burn.

Another time, I had my daughters and granddaughters at my home, and I had strung a hammock between two large trees. These are large, fifty-year-old trees with trunks at least two feet in diameter. While my daughter and granddaughter were swinging in the hammock, one of the trees just fell over, and I will always be grateful that they just fell to the ground and were not injured by the tree. I realized that I was rolling the dice with the safety gods. I am certain that everyone has had close calls. In the safety profession, we call them near misses. Every day, we hear about and see situations where people are sitting under the wrong tree. We all have a sense of overconfidence that nothing bad is going to happen to us—until it does happen.

Some of the chances people take in real life are amazing, such as people riding bicycles at night without lights or reflective clothing. Are they trying to kill themselves? I realize that they are not really thinking. They are just overconfident that nothing will happen to them. They think every driver will see them and that no driver will be drunk, tired, or texting. Young people seem to be the most overconfident. They are strong, and their reflexes are quick. I think of them skateboarding without a helmet, and those crazy drivers we see on the road, creeping up behind our car at eighty miles per hour, weaving through the traffic, changing lanes without a turn signal. Overconfidence is underrated as a cause of injuries.

IMPETUOUSNESS

To be impetuous means acting quickly, without thought or care. Some people are more impetuous, and some are more deliberate. Overall, humans make plenty of mistakes, and one mistake can cost you your life or your quality of life. In basic children's education, they teach a child to touch a hot stove, and then the child learns to never touch the hot stove again. They don't tell you that one touch of the hot stove might burn or scar your child for life. I have cases where infants and adults have been severely injured by hot water. Until those cases, I never realized how important the water heater regulator is. Nobody thinks to test the temperature of the water before putting their hands or any part of their body in danger. I learned from my cases that the water heater regulators could be put on the wrong setting or wear out over time, and suddenly, there can be scalding water.

People, by their nature, don't want to take the time to plan actions and activities. We take a lot for granted. We are in a hurry to get our actions and tasks done and move on to the next task. Impetuousness causes problems when lifting. For some reason, people don't take the time to think of the consequences of lifting heavy, awkward objects or even other people. People get seriously hurt trying to carry another person on their shoulders, and countless times, people end up with surgery after trying to lift or push a heavy object. I have seen this so many times that I am starting to believe that humans don't have a guard against impetuousness. In human evolution, people just didn't lift heavy objects and never developed a sense for when it is safe to lift or push and when it is clearly not safe. We need research in this area, but for the time being, I tell people not to lift or push. Get specialized equipment to do it or hire a professional with the right equipment. This is some of the best advice I give, but nobody cares.

IMPATIENCE

Impatience is like impetuousness. For the purposes of this description, they are the same. Impatience is more of a quality, while impetuousness is more of an action. An impatient person can be a dangerous person because impatience often takes the form of anger and sudden movement. It makes me think of all those drivers on the highway pulling into another lane without being certain it is clear. I have been stuck in a stopped lane, and the lane to the right has cars going eighty or ninety miles an hour. I should wait, but I get impatient. A driver is impatient and becomes angry when another driver cuts him off, fails to move over in the fast lane, or fails to move at all after the light has changed.

Impatient people, from my experience, are some of the most dangerous people on earth. They are always trying to get other people to hurry and not take the time to take action or do the job with preparation and planning. I try to weed out, or at least identify, impatient people. Impatience is the enemy of safety. Most tasks, especially dangerous tasks, take time to evaluate and plan. It takes time for the electrician to climb down the ladder, walk over to the source of the electricity, turn it off, and lock it out. It would be so much easier to just try to make the connection right then on the ladder and save all that time and work. Once again, human nature working against safety. That electrician has never been electrocuted before, and they think it's fine to take a little chance this time. They will try to be careful.

IMPAIRMENT

Many people believe they can drive their cars safely if they have just one drink, or smoke just one marijuana cigarette, or eat just one edible. They believe they are not over the limit, so they are legal. This is not how the human body works. Having any alcohol or drugs in your body impairs your performance to some extent, and it impairs your reaction time and judgment. The impairment varies significantly from individual to individual, and even blood tests won't tell us how people perform tasks with one drink, one marijuana cigarette, or thousands of other drugs that clearly say, "Do Not Operate Machinery." Various drugs have different effects, depending on body weight, the time they are taken, and the person's physiology. This is a complicated issue requiring scientific study, but nobody cares. A safety professional knows the human performing a task should not be impaired at all, whether it be drugs, alcohol, fatigue, distraction, or many other conditions, including poor health. It is not uncommon after an injury or a fatality for the blood tests to reveal an impairment. This is a big problem in our society, and it is getting bigger. Until the computers take control of our vehicles and equipment, a safety professional must worry about all kinds of impairment. Inexperience is another kind of impairment. I recently watched young adults try to use ladders and scaffolds for painting, and it was scary. I have also seen people using a chainsaw for the first time.

PANIC RESPONSE

Something goes wrong with the operation or the machine, and the person operating the equipment does the unexpected. They may just run away and let the danger loose. I have seen cases where people are using high-pressure water hoses, and something goes wrong. They may not have a firm hold on the hose, or the water pressure might be higher than expected, and suddenly, the hose goes flying around the room, injuring people and, in one case, killing people. High-pressure water or air can get into the skin of an unsuspecting person, and once in the bloodstream, it is almost certain death. Panic response is a terrible and dangerous thing, and it is a normal human response to fear. A panic response could be described as your body's emergency response.

The best thing a human can do is not to panic, but it is not controllable in many individuals. People have social anxiety, and when put in the right position at the right time, they act unexpectedly. Everyone has these fears. If your coworker goes down with a heart attack, will you know what to do? Will you know how to act? Will you act, or will you run away?

PHYSICAL HUMAN LIMITATIONS

Not everyone is the same height and weight. We all don't have the same strength. Some of us have significant physical limitations. A ladder might be fine for a person of a certain height, but for a taller or shorter person, the ladder might not be the right tool. The same is true of physical strength and dexterity. Not all jobs and tasks are for all people. I have investigated the root cause of an injury or fatality and found that the injured or dead person was tired or sleepy. I know of a case where a car dealer had their salespeople do snow removal after snowstorms. Having salespeople shovel snow is not a smart idea. Salespeople don't have the physical ability or equipment for snow removal. In the end, it will cost money. People who do physical jobs train for years and learn their limitations. They develop techniques and procure equipment that facilitates the success of their tasks without causing significant pain or injury. Placing a person in a situation where they have no experience and no training and haven't developed the physical capabilities necessary to perform the job is a bad idea.

ACTIONS SPEAK LOUDER THAN WORDS

The world is full of misleading information and materials. Advertisements are exaggerated, untrue, and misleading. They tell you a product is safe when it is not safe, or if it is dangerous, they hide the danger deep inside hundreds of pages of small print. There are schemes to take your money. There are knockoff products and fake products and offers we all get every day. We cannot trust that people and businesses are looking after our best interests. We cannot trust their words. We must evaluate products and services based on their actions or on our own experience.

This lack of honesty and openness is a serious problem in the safety and health field. A safety professional cannot rely on a piece of paper that says someone has a procedure or a program to protect us from danger. We need to see proof that it will work. When there is a broken stairstep, a bad ladder, an open electrical circuit, a wet floor, or falling ice, we don't need words. Somebody must take action to eliminate the hazard. Any person who fails to act to correct a hazard is a liability. It is not acceptable behavior to walk by a hazard and believe it is somebody else's problem. When individuals or management are aware of hazards and fail to take action to eliminate the hazard, it should be a crime. Unfortunately, we live in a world where very few people take the time to report a danger, and managers think that a warning is the same as a permanent fix. I see a dangerous step down from a storefront or a house, and nobody is going to do anything about it. People will continue to trip over the step again and again until somebody is seriously injured and hires an attorney—and then, and only then, will the hazard be corrected. I have been to a house where they have a two-inch drop-off between rooms, which is not illegal but a terrible danger. People who live in the house are used to the drop-off after many incidents. After a few near misses or minor injuries, they negotiate it without much difficulty. However, those of us who are not accustomed to it will be surprised by this small defect, and since we are not looking down when we are walking, we can be seriously injured. One time, I put down a towel to remind me of the drop-off, and it was promptly removed.

INSANITY POTHOLE DILEMMA

Having the same injury over and over and expecting different results is insanity. It reminds me of potholes. In the spring, I run into potholes in my car, just like everybody else. Some of these potholes are bad, and some of them will break your tire and your axel. We all run into these potholes, but nobody reports them to authorities until there is a serious accident or damage to the car. There is no systematic method of reporting and correcting pothole hazards. These potholes represent a significant danger to all of us and to our families. Nobody does anything about it until somebody pays a horrible price. Why is this the case? When there is vehicle damage or even a personal injury, the insurance company pays, and the car gets fixed. There is also a feeling that reporting these dangerous hazards is a waste of time. The whole idea of a safety program is reporting and correcting hazards. When we take the time to report hazards, the responsible parties, such as the highway authority or road authority, have more liability, and they are more likely to act and set up a program to eliminate these hazards and reduce their liability. The safety engineering solution to this problem is that potholes must be reported, identified, and repaired, and until they are repaired, they must be barricaded. We know there are not enough people to travel all the roads and highways to inspect for potholes, but we also know that cameras are cheap and everywhere, and the camera will have no trouble identifying potholes. This is not a difficult solution. Place cameras to monitor the road and send crews to at least blockade dangerous potholes until they can be repaired. I know that nobody is doing this, and I suspect nobody is going to do this, but it does point out problems we have that could be solved with technology but are not being solved. I can't believe we don't have backup cameras on all the trucks on the roads with gigantic blind spots. The problem is that the safety of the individual is not a high enough priority, and we need to turn that around. We must change the culture to make it a higher priority.

THE BEGINNING OF BECOMING A SAFETYMAN

IT STARTED WHEN I WAS GROWING UP. THIS WAS IN THE EARLY 1950s, and it was a simpler time—a time when the world was moving a lot slower, there were fewer people, and there were no computers, internet, or cell phones. We had a feeling and a belief that seems odd today—that we were safe and that the government and our neighbors would be looking after us. To some extent, it was true in the old days. People seemed pleasant, kind, honest, and generous, and we felt safe wherever we had to go, except when we went to war. We had survived the Great Depression and the World Wars, and there was an air of positivity and prosperity in America. There were hard and difficult conditions overseas, but they were so far away, and they were not in our face the way they are today with television and the information highway. We were living a *Leave It to Beaver* culture where the family unit was protective and caring, and we had *Gunsmoke* as an example of how good always wins over evil.

Marshall Dillon will take care of it all. The world was a happy place and not a cruel and dangerous place.

It may be difficult for younger people to understand that we were loose in the world, and there didn't seem to be much danger. We had freedoms that are clearly gone today. When I was six years old, my parents let me walk to school, at least a half mile, by myself, on the South Side of Chicago. I went to the same grammar school as Michelle Obama; some years before, she was there. One time, I went back to my grammar school and attempted to take pictures of the playground where I used to play, and before I could even snap one frame, undercover police officers jumped out of their car and grabbed me, and I thought I was going to be arrested. That event at my old grammar school tells the whole story. Things had really changed; where I had innocently played was now a dangerous and protected place. Someone taking photographs at a children's playground is very suspicious. It is a good thing that they let me go.

In the sixth grade, my parents decided that we needed to move to the suburbs. The suburbs had a different culture than the city, and this change was not good for my self-esteem. Many people dream of growing up in the burbs and going to a well-known high school like New Trier High School, but for me, it was just overwhelming. I think more than anything else, it was just a big change at such a sensitive time in my life. Looking back, I must acknowledge that I am grateful to my mother for moving us to the suburbs. That move, with all its unhappiness for me, allowed me to get into the University of Wisconsin with just average grades from New Trier High School. It was at the University of Wisconsin that I found my safety standards, safety programs, and wonderful teachers. The university gave me the opportunity to think I could make a difference in this world.

My safetyman career started a little earlier than college. It started on a day I can remember as clearly today as the day it happened. On that day, one of my friends sustained a serious head injury while sledding. It was in 1959, and it was Christmas vacation. My friend and I were bored and looking for excitement. There wasn't much excitement for preteens in the late 1950s, and we decided to go sledding. One of the other activities we enjoyed was swimming in an old pit called Ditches Pit. It is hard for the safetyman to believe that we would swim in an old, still water pit. There may have been chemicals in that pit from dumping, but they couldn't have been too toxic, or we would have been long gone by now. One time, we were swimming, and we saw a rat swimming with us, and that was the end of Ditches Pit. Anyway, on this occasion, it was sledding, and we needed a hill. This was sledding like Rosebud sledding; the kind of sled I had was like the one from the movie *Citizen Kane*.

I had that old wooden sled, and he had a metal one. Where we lived, there weren't too many hills for sledding. We decided (I think it was my friend's idea) to go sledding near an embankment on the highway near a shopping plaza, where there were hills forming the embankment of the new overhead highway. This hill was just a short, gentle slope and not much fun after a few runs. That's when we decided (not sure who, it could have been me) to try sledding this short run backward. This meant on our backs, looking at the sky, with our heads facing downhill. This seemed like more fun, and I was the first to go down the hill. When I got to the bottom of the hill, I climbed off my sled and was dusting off my jacket, and about fifty yards away, I could see his sled was veering to the right (my right), and he was heading for a small tree. I had no idea there was any evil in the world or that anyone could get hurt. I do remember a friend Larry ran into a metal post in the alley behind an apartment building where we would play touch football, and he always walked with a limp. I also remember we used to play inside cardboard boxes like refrigerator boxes in that same alley, and I think we were fortunate to never have been run over by a car.

Besides that one-touch football experience, I don't think I had ever seen anyone get hurt aside from some cuts and scratches. I saw his sled hit the tree, and I thought he might have a bump or a scratch. To my horror, after the hit, I saw that he was not moving. I ran up the hill as fast as I could, and as I got close to him, I saw something on the hill next to his head. I could not believe my eyes because what I saw was his ear; it had been torn from his head and was hanging by a string of flesh. I got closer, and I could see that his head was cracked, and a strange liquid was running from his skull.

I quickly grabbed my good friend and carried him up to the top of the hill. I tried to stop a few cars (no cell phones then), but nobody would stop. Finally, because this was an actual emergency, I jumped in front of a car, and that car hit my leg. To this day, I can still feel where that car hit my leg, at least in my brain. I got my friend into the car and to the hospital. Six long brain surgery hours later, I learned that he was going to live. As far as I know, my friend is still alive to this day. They reattached his ear, and he looked pretty good for six hours of surgery, and I was proud of what I had done in an emergency.

The only problem was that my friend was changed forever; his face was changed, and his speech was changed. My friend was not the same person. Nobody would even say the horrible words—*brain injury*. Nobody had a brain injury! Well, my friend had it, and it wasn't pretty. I was so scared, and I felt bad, not like a hero at all. I had now been involved in a terrible situation, and I had no idea what to do next. I didn't have any coping skills. I went home and told my parents and brothers.

Even though my friend had suffered this terrible event, our life was unchanged. They told me that there was nothing I could do and everything would be fine. I somehow knew at that young age that things were not all right and that they were never going to be the same for my friend and, to some extent, me.

I went to see him at his home. His parents and his sister didn't look too good. My friend was going to get better, but I think we all knew he was never going to be the same. When I was at that house for the first time, I got a feeling that has never left me to this day. The feeling was that his family thought it was my fault.

This feeling was life changing, and there was no consoling or counseling that was going to change this fact. There was nothing to do except to blame me and for me to blame myself. My family did not blame me. My friend's family blamed me. I could see it in their eyes. They blamed me fully and completely. If he had not been playing with me, this would never have happened. I never did find anybody who cared about my feelings after doing the right thing and then being blamed for no rational reason.

This was the ruining of my life. It drilled down into my heart and soul, as a young person who somehow lived through the ordeal with survivor's guilt. Survivor's guilt is a mental condition that occurs when a person believes they have done something wrong by surviving a traumatic event when others do not survive. This term was first introduced in 1961 by William Niederland, a psychoanalyst, as a way of describing the experiences of Holocaust survivors. Symptoms of holocaust survivors guilt included having depression; flashbacks; feeling irritable; having difficulty sleeping; feeling immobilized, numb, or disconnected; being unmotivated; feeling helpless; experiencing physical symptoms, such as headaches, stomachaches, and palpitations; and even having suicidal feelings. I had all of these, and then I had to return to school. "What happened to your friend?" I wasn't too popular without my friend in the first place, and now I was not popular at all. My grades suffered, and I didn't want to be at school. I wanted to die, and I never talked to a single person about my feelings. Mom was wondering why things were changing with me. Nobody had a clue that the horrible incident that happened on the hill near the highway changed and shaped my life forever. I was totally unhappy.

Somehow, I got through high school and went to college at the University of Wisconsin. I had no desire to go to college at all. I just wanted to get away from my own feelings. I realize that deep-seated problems can be secret things that nobody wants to talk about, and the pain can last for a lifetime. Sometimes, people think I could be a little happier and more positive and smiley. I think this must be genetic because my father, who was a quite pleasant person, was given the same criticism. I know many people who laugh and smile a great deal, but I am not certain they are any happier than the rest of us. I can't help or change who I am, but they don't know what it was like for me in those dark times that have never left me.

Things started to get better in college. It was a different environment—new people and a fresh start, and college seemed easier than high school. I joined a fraternity, and I wanted to learn more about myself and other people, so I took every course I could in psychology. I know now that I was trying to get tools to survive in my life the best I could. I found courses in industrial psychology, taught by Dr. Karl Smith, the most interesting. He taught about behavioral cybernetics that were orientated toward science and made the world feel more stable to me. No matter what I say about Dr. Smith, it will not do justice to what he did for me. I could tell he liked me from the start, and he thought I was smart. When I read his papers and his books, everything he said about how people work and learn was consistent with my beliefs. I did so well with Dr. Smith because it didn't seem like work at all; it was more like we were kindred spirits, and I instinctively understood what he was saying about computers, statistics, and methodology for testing theories and hypotheses. Working with him and his fellow graduate students in the old cybernetics laboratory beneath Rennebohm Drugstore on University Avenue changed my life and sent me on a mission of success and satisfaction that I enjoy to this day.

SAFETYMAN PRINCIPLES AND RULES

THE PRINCIPLES OF BEING A SAFETYMAN CAN BE THE SAME principles for living a good, healthy life. Following certain rules and guidelines have helped me be consistent and successful in my field. I must be objective and consistent in my work. Rules are based on the belief that life and safety and health protections are required by our society and our laws. When someone is injured or killed because of a defect or the actions or inaction of another party, they are entitled to compensation, and a corrective remedy must be made to assure that this same defect or action does not happen again. Repeated cases of violations of the same safety principles should be considered crimes. When a company violates safety rules and a person is injured or killed, and later, the same violation occurs at the same company under the same management, it should be considered a serious offense. When an individual without proper fall protection falls from an elevation, and then later, another person falls for the same reason—no fall protection—this is like an intentional act and unacceptable. Some states have legislation called intentional torts that make it a crime to violate safety and health rules and regulations when it can be proved that there was prior knowledge. Failure to act in a reasonable manner to prevent a reoccurrence of a serious hazard must be a crime.

PRINCIPLE NO. 1: LIFE IS WORTH SAVING— PAIN IS WORTH PREVENTING

You must have a heart. Some people are so cold and don't care about the loss of life or pain and suffering. It is narcissistic behavior—people who only care about themselves. People get to be any way they want, and there is plenty of room for selfish people, but most of us caring people must outvote them and move the ball forward toward a better place for all of us in this world. I have learned not to get angry at selfish people, and I don't try to convince them to change their beliefs. I try to get them to work with me for some benefit for their own situation. Most of the time, if I can get them to believe it will put money in their pockets, I can get some help.

PRINCIPLE NO. 2: MONEY IS WORTH SAVING

Not everybody is motivated by money, but most people understand the value of having or saving some money. The savings from preventing accidents and injuries is so substantial that my job is to make sure that the individual knows that this is not just money that goes into some government coffer but the money that the individual can actually put into his or her own pockets. People need to know that by supporting the safety and health program, they will get some of the benefits, or they won't care at all.

RULE NO. 1: FOLLOW THE MONEY

We live in a capitalistic society, and money is power. Following the money is important. People who are paying for things have a great deal of power, and they are able to decide the quality of the safety and health effort on a construction site, at a factory, and in our environment. We must have a message and culture, starting from the top, to make the biggest impact on cost savings. The idea of having top managers appreciate the value of standing up and taking responsibility for the safety of the entire operation is contrary to contracts and management personnel who want to send safety and health responsibility to the lowest level. They say, "Everybody is responsible for safety and health."

The problem is that not everyone is able to establish a culture for safety and health. Many companies and managers have told me that they believe each contractor, or each person, is responsible for their own safety. I do not believe this to be true. It starts at the top, and it is the people in control who determine the means and methods for the overall safety effort. When the contractors and workers know that only safe behavior is accepted, they respond in a positive manner. To get safety and health, we must make an impact on the culture to allow a safety professional to be successful.

This same idea of following the money is used to determine where the most money will be saved. Safety culture and safety management require good management, and by implementing a safety and health program, there should be cost savings. When I work for a client, I must show them I am worth the expense and that safety does pay. I look for where money is being spent on injuries and illness and direct the safety and health program in the direction of saving money. I want to make certain it is being spent where it can do the most good. This is the most potent weapon there is. One of the best things about being a safetyman is I can change the culture and save human suffering and money at the same time.

All expenditures must be justified. Following the money is just a method of prioritizing my time and efforts to get the best results. My clients must see my work as a profit center affecting the bottom line. Finding the best way to spend my time and my client's money allows me to figure out what tasks I need to complete each day and in which order.

RULE NO. 2: NEVER DO ANYTHING IMPROPER, ILLEGAL, OR IMMORAL

Setting a safety culture means making the safety professional an example for everybody who is not as focused on safety. Being a safetyman means just what it says. I am a man of safety. Nobody is perfect, but I do everything I can to set an example for other people. If I am perceived to be a hypocrite, then nothing will change, and I have no right to expect people to listen to me or follow my example. This is not limited to safety and health. I must have leadership qualities to get people to listen and understand what we need to accomplish and why it will benefit the entire organization and the individuals involved. It's beyond safety. It has to do with what kind of person I am and what people think of me. I need the people to respect me and my effort if there is any chance of initiating a cultural change. My clients and the people I work with must see me as an inspiration for safety and morale. I believe that safetymen and safety women who exhibit traits of nonconformity make the job harder to accomplish. When I see a safetyman or a safety woman who is smoking or riding a motorcycle without a helmet, I see a problem with their leadership. If the safety people can't get the attention and cooperation of all levels of management and workers, it is likely that they will fail and fall victim to the temptations of danger.

RULE NO. 3: GO TOWARD TROUBLE

I always go toward any trouble when I find it. Problems are the friend of a safetyman. Wherever something is not going well or there is some conflict, there is a time for the safetyman to intervene and prevent a problem from growing into a major catastrophe. I never walk away from trouble; I try to face it and deal with it. Most people want to get away from trouble, and they just walk away, but that does not work for the safetyman. Many people have an aversion to trouble. These people don't like conflict and will go to great lengths to avoid a negative interaction with other people. A safetyman must deal with trouble immediately, or it will come back at him. Enjoying conflict must be one of the most important traits of a safetyman. Being a safetyman in this world has meant constant conflict with other people who do not support a safety culture or who don't want to be told what to do. I can think of very few situations over the many years that I have worked in the industry where there was not some kind of conflict. Conflict is the way of life for a safety professional because they are always balancing the pressures of production and individual freedom. People have told me that they just don't care to wear safety glasses or a hard hat. Their faces tell me they think I am imposing on their individual freedom to wear what they please. This does not work with the safety professional or the safety culture. We cannot afford to allow everybody to do what they wish. It is too expensive to pay for easily preventable and unnecessary injuries and illnesses. A safety professional had conflict every time they try to impose their will on management or individuals, even when it is clearly for the benefit of the entire organization. A safety professional learns to go toward the trouble, and they better learn to accept this conflict and maybe even enjoy it a little bit. They say no pain is no gain.

RULE NO. 4: NEVER ARGUE ABOUT MONEY

It's a lose-lose situation to fight with clients about money. If they don't believe they are getting a benefit from my services and from safety and health in general, then I am not doing a good job, and I might as well go somewhere else. Not everyone wants to live in the world of a safety professional. Some companies and people want to hold on to the old ways of doing things, or they don't care to change their ways. I have encountered companies that would rather lose money than implement safety and health programs. If they can't see my vision and they don't want to come on board the safety train, all I can do is go somewhere else where I can be more effective. There is no reason for me to argue about payment for my services when I already know my services will not do any good. My clients tell me how much they are on board with safety and how much I am worth when they pay me. I never argue or fight over money. I have made a decent living, and at the end of it all, my work is not about the money. I know I will be paid when my clients see results from our combined efforts. I focus on doing the mission of changing the culture and eliminating the hazards, and the money takes care of itself. When safety becomes a profit center, my clients think my fees are small.

A SAFETYMAN SHOULD BE A GOOD GUY

I AM A SAFETYMAN, AND I CAME FROM OSHA. WHEN I TALK TO people, the reaction is often negative. They think I am an enemy of the company or the country. They think I am trying to take away their freedom. I have had people roll their eyes and walk in the other direction, becoming angry when I tell them the safety glasses are mandatory. Many people want nothing to do with OSHA or safety. They think the principles that make up the basis of my life are going to hurt their individual desires or freedoms. This doesn't make any sense. People think regulations and standards represent an impingement or restriction. They believe that reducing regulations and standards means more freedom and profits and a better economy. This is their attitude until they face a devastating injury.

Many people believe that their individual freedom is more important than safety and health. These people have every right to believe whatever they wish, but I have seen those beliefs change when they or someone they know suffers a devastating injury or illness that was no fault of their own. I tell them that I too believe in individual freedom, but part of the freedom is for me to have information to take action to protect myself and others from catastrophic losses. I ask them how they would feel if they lost an eye when they could have been protected with safety glasses. Sometimes they put on the glasses.

Putting the individual freedom concept ahead of protecting your body is wrong and costly to so many injured people and families. People must know that there is nothing more important on the earth than the safety and health of the people. It is not possible to have a healthy country, healthy companies, or a healthy economy without healthy people. This should be easy to understand. There is constant rhetoric regarding "unnecessary" regulations. How can a regulation designed to protect our people be unnecessary? Until a defective product hurts your family or some agency doesn't act to protect your people, it's difficult to get people to understand that this does not always happen to somebody else. There is no more important function of government and society than protecting our people. Those of us writing and enforcing safety and health standards do not have a political agenda. We come from every type of background, ethnicity, and political affiliation. Our only agenda is to protect people from well-understood safety and health hazards.

Those of us engaged in trying to make certain that companies and products don't violate safety standards or have defects are the good guys. Most of us are businesspeople, and if we don't show our clients how safety and health will make them more productive and save them money, we do not get any work. Our job is the opposite of a political agenda. We are trying to make safety and health profitable for every business. I believe I can show any business person or company that does not have a fully developed safety and health program that their safety unit can be a profit center. I have told many potential clients that I would work for a percentage of the savings of the implemented safety program.

Safety people try to improve the culture, morale, productivity, and bottom line for every company. We understand how difficult it is to make a business succeed. Nobody wants to fill out unending government forms and surveys. We sympathize with owners and managers who want to get to the work and avoid the government bureaucracy. They want to lay concrete or dig a trench or paint a house, but they must deal with local codes, all kinds of inspections, evaluations, and OSHA. This is one of my selling points for safety and health-consulting services. We tell these owners and managers that we will get them through those forms, and we focus on examining areas of economic losses that can be turned into productivity and profits. Every company without a fully implemented safety and health program has significant losses. Workers' compensation, downtime, medical costs, hiring replacement workers, first aid, hospitalization, and litigations are areas of potential saving with an improved management effort. Improvements in culture and climate can be the lifeblood of a growing company.

A safety professional should represent good management. Safety management is good management. Workers who know that the company cares about them, their families, and their lives are more enthusiastic and energized in their work. Safety people are the friends of the business. We support their success, and we try to improve their bottom line. A safety professional should be viewed as a friend of the business because we are preventing catastrophic losses and upgrading and improving management that will benefit the entire operation. I appreciate that nobody wants to see the OSHA inspector looking for violations and giving out financial penalties. But the mere presence of

OSHA has a positive effect. It makes everybody think about safety and realize that there might be hazards that need to be corrected or that have been overlooked. That OSHA visit is a free consultation. Many companies have hired my firm to do OSHA-style audits, and they pay thousands a day for our professional work. When the OSHA compliance officer arrives, most likely, the inspection will result in a small penalty much less than the cost of a private consultant. I tell my clients they are getting a good deal, and if they do get in trouble, OSHA will be reasonable if they make the necessary corrections.

This negative attitude about safety is one of the biggest problems in the world of the safety professional. I wish I could get every business person and company owner to think about it in a positive way. Where is the negative? What is the downside of having an upgraded safety and health program? I believe that many of these company owners are being shortsighted in that they are just looking at the initial cost of having a safety professional or an OSHA visit. They need to step back and look at the long-term benefits. One of the most positive changes I have seen over the years is the change in attitude among companies that have dealt with safety problems. Once they deal with a safety professional or with OSHA, they begin to see the benefits. It is not going to be the safety professional or OSHA that will cost them big money but the injury or death that will lead to so many bad things, including litigation.

The safety field is an enormous source of good jobs. You don't have to go to a prestigious university to become a safety professional and have a good wage. Safety professionals can come from the building trades. They can be welders, painters, or laborers, and they can specialize in the area where they have had the most experience. As an example, a laborer spending their time learning the proper way to lift and move materials is a valuable resource. That laborer will have learned to work and use the equipment for material handling that will not damage their body. This laborer has become an expert in safe lifting and material handling. That laborer can get some training at OSHA, take a few college courses, and became a safety professional. Becoming the safety representative for their company or their trade is a road to advancement. I have known of many people from the trades who have started their own safety and health-consulting firms.

The people working in the factories and construction sites are realizing the potential for upward mobility in the safety and health profession. I get calls and emails every week from people working in the trades who want to become a safetyman or a safety woman for higher pay and a better life. They are on the right track, and there are tremendous needs and a demand for competent and knowledgeable safety people. I suggest they take some courses at OSHA and try to get a job at OSHA because that is where you get the information and experience that can make the American dream come true. I believe that employment in the safety and health fields and many specialized fields related to safety and health will increase exponentially in the next decade. I see colleges having more comprehensive curriculum in the safety and health area, and it is exciting. An interested person could pick almost any area from the standards and become a specialized expert in that area—taking trenching and soil mechanics as an example. An interested person could spend a lifetime learning about soils, the effects of water, the testing, and the equipment and before long be offering services to the industry that will be most welcomed. Safetymen and safety women will make our country strong and vibrant, and they will become upwardly mobile.

MEN AND WOMEN IN SAFETY

IN MY FIRST YEARS AS A SAFETYMAN, THERE WERE NO WOMEN and very few minorities. Today, they are some of the best safety professionals. It is a great advantage to speak Spanish. It used to be a terrible chauvinistic environment for women and minorities on the job. Today it is better, but we still could make room for more respect and upward mobility in the safety field. I think the more women and minorities that enter the health profession, the better off we will be, especially since the workforce now includes a much higher percentage of women and has always had a high percentage of minority workers. Different people from different cultures and men and women have different safety and health issues and different approaches to safety and health problems. There are many opportunities for those groups that have been historically disadvantaged in the safety and health profession.

I try to teach these potential safety professionals that the key is to show top management how they can save money. They should offer to work on a percentage of the savings in workers' compensation costs. If any business is sustaining injuries or illnesses on a regular basis, a safety and health program will pay more than one safety professional salary. If a company has minority or female workers, and nobody is there to speak to them about their issues, a person with an understanding of their problems can be a fine addition to management staff.

I have often spoken of culture to management officials who have never taken the time to think about the condition of their workers. A culture where management cares about their workers is fertile for increased productivity and safety. I have seen situations where managers and workers don't even speak the same language. There is no more valuable resources for finding problems that might cause injuries and illnesses than to talk to the workers themselves. I have conducted many accident investigations where there is something wrong with the plant or the construction site, and I can't put my finger on what the issue is that's causing the workers to act timid or have low morale. Sometimes I see that people are afraid to talk to me. This was especially the case when I worked for OSHA. If the people are afraid to talk to me or to OSHA, right there, I can see there is a problem in the culture for safety and the culture in general. A situation where the workers are afraid to speak their mind is always troublesome. To overcome this problem of lack of communication, I will talk to the workers in private or get their telephone numbers and call them at home. When they feel safe, I can find out things that must change, and when I make those changes, it is better for everyone.

SAFETY WOMEN

Not enough women are in safety. You could say safety is a man's world, but it doesn't have to be. I do think that safety can be intimidating to some women in that you have to deal with the working men, and they can be a little crass. Now, I don't think I can speak for women or even be a defender of women, but I have heard the wolf whistles, and I have heard the locker-room stuff. I wouldn't like to have my daughters exposed to that stuff. These things are changing fast, and the Me Too movement is making me and many other men feel the change. Today's women have no reason to feel intimidated by the man's world. I hope this is the case because the world of safety could use a lot more women.

INTERVIEWING FOR SAFETY

When people are afraid or just not interested in talking to someone from the outside, it is an opportunity for the safety professional. When workers are afraid, unhappy, or seem uninterested, there is much to be learned. We can evaluate the injury and illness logs and see an opportunity to make things better. Some of the most difficult duties for the safety professional involve interviewing the workers in private and getting and having a meeting with top management, also in private. I mean *private*! These are very sensitive situations, and if the interviewer or investigator is not sensitive to the situation, it can blow up in their face.

One time, there had been two fatalities at the same plant. I was with OSHA, and the workers I was interviewing were scared. I was trying to find out about a supervisor who was involved in both cases and who was reported to be mean and unsupportive of safety issues or even basic kindness. I asked too many questions about this supervisor, and one of my interviewees told the management that I had something against this supervisor. It came back to me that I was trying to get this guy. That was not true, but I did find that he was putting fear into his crew. They would not report safety and health issues and felt they could not speak up about safety. When something like that happens, the best thing for me, as a safetyman, to do is to have a conference with the owner or top manager and try to get them to understand how important a culture of kindness and safety is to the business in general and to the safety and health effort. Sometimes I had the feeling that the manager was listening, and things were really going to change, and other times, I got the feeling that what I was saying was going in one ear and out the other. In the end, my success or failure is based on the relationship I have with this top manager. They have confidence in me and what I am trying to accomplish, or they just can't wait for me to leave. Confidence is so important when trying to change cultures and save lives. Safetymen and women must have the confidence to stand and deliver with the biggest boss and the most fearful worker.

THE SAFETYMAN CAREER

I NEVER KNEW THAT BEING A SAFETYMAN WAS GOING TO BE my career. I was interested in trying to do something positive in the world, but when it led to safety and health, I found it was not a high-level position. Many people have never thought much about safety and health career. Community colleges and universities have only recently started developing curriculum for a degree program in the safety and health field. It is such a broad field, almost like the medical profession, where you can be a generalist, but you can also become a specialist, and there is so much potential for upward mobility. If I had a choice between law or safety and health, I would still choose safety and health because there is less competition.

In the years I have been working in the safety and health field, there are few people I have encountered who have the background to practice successfully in my field. This does not mean that the opportunity is not there. It is just that there are few colleges or institutions that offer the safety and health degree program. I can envision the need for hundreds of thousands of safetymen and safety women in every industry. I have known individuals who have specialized in trenching and soils or electricity that have gone on to have successful careers. One of the leading people in the safety and health field has specialized in fall protection, and I know of one person who specialized in personnel and debris nets who sold his business for millions of dollars. An interest in the safety and health profession can go a long way, and the compensation for positions in safety and health exceed the compensation of many other related fields, such as teaching and nursing.

A few times a week, we get telephone calls from all over the country, and sometimes around the world, from young people working in a factory or on a construction site who have discovered the safetyman website and want to become a safetyman or safety woman like me. It must look pretty good to them from the desire I hear in their voices. Usually, they want to be certified, or they want to get OSHA approved, but I think they want more money and a better life for their family. I tell them that although nobody can certify them or get them OSHA approved, they are on the right track, and the fact they have called me tells me they are motivated in the right direction. I tell them about my experience of staying close to OSHA and trying to make friends at OSHA that eventually led to a job as a compliance officer. That job opens the doors to so many opportunities. Of course, not everybody can work for OSHA, but if they use the internet, contact the nearest OSHA Education Center, or go to the OSHA website and look for the educational experience, they will be on their way to a new and exciting career. No matter where you live, there is safety education available at OSHA, from the unions, or at the local community college. I promise them if they continue to build up their résumé and get close to OSHA, they will be on their way to upward mobility.

I believe this is true. I know of several young people who were unstoppable in their pursuit of safety and health, and all of them have become more successful. It takes tenacity, but anyone with enough desire can have a better life through this wonderful career.

WHAT DOES A SAFETYMAN ACTUALLY DO?

MY JOB HAS BEEN TO GO WHERE OTHERS HAVE BEEN MAIMED or killed and determine the cause and origin. Sometimes there is a blatant violation of standards, customs, and practices, and sometimes we must do an investigation or even a reconstruction. Recently, I had a case where a worker was killed when he was crushed between a forklift and a dock plate. Since the forklift was stationary, this case did not make any sense until we reconstructed the incident and saw the slight drop-off that caused the forklift to move just enough to crush this worker who left the machine to look at the material on the dock. The wheels of the forklift were balanced upon this half-inch drop-off, and that slight change in elevation was enough to crush his body, which was squeezed in between the forks and the dock.

When you read about an explosion, crane failure, or trench collapse, or when someone falls into a machine, that is where I go. It is interesting work going toward the danger and investigating the facts.

I go into confined spaces where people have died with no air or where they are overcome with chemical vapors. I have been to so many dangerous locations in a forty-six-year career. These places include the steel mills, mines, thousands of factories (big and small), railroads, and major construction projects. I was at the Boston Harbor "Big Dig" and the expansion of McCormick Place in Chicago, United Center building, the expansion of Terminal No. 5 at O'Hare, the tunnel between the strip and the airport in Las Vegas, the tunnels in Hawaii and Milwaukee, and thousands of other places. I have learned so much in these places and have had the opportunity to teach courses and write standards about the danger. I have seen so many violations of standards, customs, and practices. Sometimes I see things that scare me and, in many instances, situations we call imminent danger or immediately dangerous to life and health (IDLH). I try to talk to people and get things fixed before something happens, but I am not certain you can imagine the reaction when telling people about danger or about what is wrong with their job. They *never* thank me; they are always angry, and 90 percent of the time, they won't change the situation to make it safer or better.

I tell the guy who is bringing the compressed gas cylinders into my health club that the cylinders need to be secured, and they need valve protection caps, or they could become torpedoes in the club. I have seen torpedoes before, and the guy with the cart of compressed gas cylinders tells me to mind my own business and not tell him how to do his job. I carry pictures of cut hazards, pinch points, and mold in my camera at my club for anyone to see, but nothing ever changes. I have tried to explain to management the importance of planning. I use a conference table to ask them if we should just pick it up, or maybe we should weigh it and measure it and make sure it will fit through the door. I have said a million times (not an exaggeration) that if management does not have specific work rules, then communicate the rules to all parties, monitor to make sure rules are followed, and enforce when rules are not followed, they have no safety program. Most places have no safety program, and nobody cares. This is a book about safety but not about forcing people to comply with the law. Talking about getting people to comply would be too boring, and nobody would read it because they don't care.

I believe that lawyers care about representing their clients and winning their cases. I believe juries care as they are put into the unusual position where they must decide what is right for the human condition and for their own families. I have made a nice living by talking to individuals and to juries about safety, and when I see that look of acknowledgment that somebody should care about other people, it gives me hope.

The most important job I have as a safetyman is to get people to trust me. They must trust me enough to understand the importance of establishing a culture where everybody is working as a team, trying to take steps for the prevention of injuries and illnesses. When I go to industrial plants or construction sites, where the attitude is hopeless and where they believe there are always going to be injuries, this is where I know I am needed the most. Productive employees are proud of their work, enjoy their jobs, and never ever prefer to stay home and get a paycheck.

Wherever the culture is negative on safety or morale is low, there is work for a safetyman to turn it around. I will interview the people and see why they do not like their jobs or do not want to come to work. I believe that people should experience joy and a healthy self-esteem when they work, and they should get a paycheck that allows them to feel like they are good providers for their families. If people hate their work, hate their pay, and hate coming to work, things are going to have to change, and I believe a safety professional can help make that happen.

Many companies have human resources (HR). I have found that too often, these HR operations are just processing people, hiring and firing, and maybe drug testing. I tell the top management that HR needs to be more than a gate. The HR must be establishing and evaluating the culture of the company for productivity and safety. When I find that the HR department isn't interviewing workers and management or has not acted for the development of a safety and health program, I think it is a terrible waste of time and money. A safety professional should take a good look at these HR departments and make sure they have the right people with the proper skills to steer the culture of the company in the right direction.

THE SAFETY PROBLEM IS A DIFFICULT PROBLEM

PEOPLE PROBLEMS ARE ALWAYS THE MOST DIFFICULT. PEOPLE have their strengths and weaknesses. Some people's problems have to do with the people you have, and others are just about people in general. This is a more important question than you might think. From a safetyman's point of view, I never feel that I am stuck with the people who currently exist at a company or construction site. Sometimes personnel must be removed or replaced. After all, they have asked for or received my help for a reason, and that reason is usually that something is wrong with the culture and the people. Some people are more caring than others, you may have noticed. Some people want to enjoy their work and put a positive spin on their working life, and others just can't wait to get out of work to have a smoke, a drink, or even a fine ride on a motorcycle with no helmet. Some people want to jump out of airplanes, fly on zip lines, and have more danger in their lives. As a safetyman, I must deal with all of them, and I like to think

I can suggest changes in personnel and employment that can make a significant difference to the safety culture.

When I try to institute change and it is not accepted, or when there is pushback against changes in a culture for a more productive and safer atmosphere, I know I have personnel problems. A simple safety idea like getting people to wear safety gear can show me if I am getting cooperation or resentment. It is not always possible to protect our individual rights and freedoms and ensure safety and health at the same time. I need team players who are willing to subject themselves to some safety and health rules for the success of the organization. We all care about liberty and personal space but not to the extent that it sends our company and jobs into bankruptcy. Plenty of companies have gone bankrupt because of poor culture and injuries and illnesses. Having an unhappy workforce and accidents and injuries are sure signs that something is wrong and needs to be changed. This conflict between individual liberty and the culture of the company is where a safety professional can make the biggest difference and the most progress for change.

There was a problem right from the beginning in 1970 when the William-Steiger Occupational Safety and Health Act, promulgated by the Nixon administration, had the effect of giving the large contributing business like the National Association of Home Builders and the oil and gas industry an advantage over their smaller and middle-sized competitors. The weight of these new regulations fell mostly on the small and middle-sized industries. Hundreds of thousands of regulations from a variety of sources were put under the OSHA Act, and compliances cost millions. Despite its regulatory burden, OSHA was a big improvement in an era where it was part of the job to lose your health or to become injured. Workers were loyal to the companies, and companies and unions were powerful in the lives of Americans.

Today, it is difficult for young Americans to realize that you could get a job with General Motors, or in the coal mines, or for DuPont, and your family would be taken care of for the rest of your life. You shopped at the company store, and even at your funeral, there would be a company representative with a check for funeral expenses. The company was your friend, but you had to sacrifice a lot, maybe your health or your life. OSHA, MSHA, and EPA all started out as really good ideas that would benefit our society, but just like a lot of other things, they began to get political, and before long, it was not about protecting our safety and health but a battle between regulation and civil liberties.

In the past, losing your health or your safety was not considered a big problem in the USA. Coal miners were expected to get black lung; asbestos workers were expected to get asbestoses; and for the rest of us, it was a sort of free-for-all on a never-ending parade of chemicals and toxins developed by large chemical companies that had corporate responsibilities to their owners and stockholders. However, some problems developed that were so bad that they got serious media attention. One of the biggest was written about in 1979 at Love's Canal near Niagara Falls in New York. This was a project by developer William T. Love. Mr. Love felt that by digging a short canal between the upper and lower Niagara Rivers, power could be generated cheaply to fuel industry and homes for his proposed model city, a neighborhood of seventy acres of landfill. Young people, many just returned from military service, were sold the American dream in the form of these cheap homes, but they were never told that these homes were built over a toxic waste dump. The canal was covered by the Hooker Chemical Company in 1953 and sold to the city for one dollar, but in the late fifties, the homes were built, and it became a nightmare with trees and gardens turning black and dying and puddles of noxious substances. It was reported that everywhere, the air had a faint choking smell and that children returned from play with burns on their hands and faces. Then came the birth defects, miscarriages, and cancers caused by benzene detected in high concentrations. The owners and developers pretended it was a big surprise, and it captured the attention of a nation that did not want their families raised with the prospect of tumors and cancer.

Next, there were some drugs that went bad. There was the case of thalidomide, the story that broke in July 2009. Thalidomide had been widely used in the late 1950s and early 1960s. This was morning sickness medicine for pregnant women. It turned out that the women who took this drug had babies born with significant birth defects, including many born without arms and legs. Television was becoming more commonplace in the homes across the nation and the world, and the people could see that these babies were being born without arms and legs. People started to say, "Maybe we can't trust these companies," and even scarier, "Can we trust our government?"

Well, the government can't have people thinking they are going to let their children be mutilated by drugs or that the Gulf of Mexico is going to be destroyed by a hurricane or by oil companies. They needed to do something to make the people think that they would be protected from bad chemicals and other dangers to their loved ones. They were implementing new programs like OSHA and EPA when on December 2 and 3, 1984, there was a gas leak incident at the Union Carbide India Limited pesticide plant, which is now considered to be the world's worst industrial disaster. Methyl isocyanate leaked out of the plant and killed three thousand people and permanently disabled fifty thousand people. It is estimated that fifteen thousand people died subsequently from exposure to the poisonous gas. This incident scared everyone in the world, but it was in India.

Then on August 12, 1985, it happened in America, in a town called Institute, West Virginia, where a toxic cloud of the same methyl isocyanate caused twenty-eight people to be admitted to hospitals, and at least 135 were treated for eye, throat, and lung irritation. American companies were now releasing bad chemicals, and American people were exposed. The safety professionals of the world started to look at other places where something like this could happen, and there were too many chemical plants located close to populated areas. One of the results was OSHA's Process Safety Management standard, which I believe is one of the best standards ever written to protect the American people, and in fact, there has not been a serious release or incident since that time. There is still doubt about whether the problem is solved and what we need to do to make certain that an American population is not wiped out by toxic chemicals. Many improvements have been made in technology, processes, and standards, but there is much more work to be accomplished. These chemicals continue to be developed and are toxic and dangerous to our communities.

The initial plan to deal with the fear of losing thousands of Americans in a toxic event was to build a fairly large government agency, such as EPA, and smaller but still powerful agencies, such as OSHA, NIOSH, MSHA (Mine Safety and Health Administration), CPSC (Consumer Product Safety Commission), and several others. This set up a battle for power between management and unions and, maybe most importantly, a battle between the federal government and state government. This battle between federal and state was the most interesting in that it was really about who was going to hand out the contracts and hire the contractors that meant jobs and an improved economy. There are many stories I have heard where programs were used to gain federal or state control of OSHA in states like Indiana and California. In the very beginning, it was the federal government that was the stronger of the two, and eventually, it was the states that exercised more control over funding, jobs, and the effectiveness of safety and health programs. Some had better efforts, like California and Michigan, and some had lesser efforts, like Indiana and West Virginia.

Of course, the problem with this infighting was that none of it was about safety and health. It became a battle for political and economic power. True interest in safety and health would involve finding out where people are getting hurt and taking appropriate action. In the forty years since the inception of OSHA, not much has changed. There is some fatality data and some lost time injury data kept at the Bureau of Labor Statistics, but over the years, I have not seen a sophisticated effort to set priorities for safety and health goals and objectives. This is hard for me because, at the beginning of my career, I worked in Wisconsin where they had such a strong program and effective safety and health effort. I worked for the Board of Workers' Compensation in an area called Safety and Buildings. We had people analyzing the workers' compensation data to determine the exact locations within the state of Wisconsin where people were injured or killed, and we would send a team of safety professionals out to those locations to solve the problem and reduce the numbers. I had the experience of going to some of the biggest companies in Wisconsin to work on getting results.

If OSHA and the other agencies can't accomplish the purpose of the agency, what can we do? A good idea might be to at least investigate the accidents that occur and then take appropriate action. A fine example of this is the National Transportation Safety Board. If you have a crash, they will be there working to find the answers. You would think that OSHA would do the same thing, but no, they are spending a lot of time with "general" inspections and "recognition programs" like the Voluntary Protection Program. There has always been a battle at OSHA between enforcement and recognition programs. The enforcement people want the rules enforced and bad actors punished, and the recognition people think the industries want to do the right thing and should be rewarded for their efforts. My experience tells me that the safety problem is so difficult to accomplish with all the payroll and production pressures that you will not get compliance or reduce injuries and illnesses unless you investigate, inspect, and enforce. I do think recognition programs are good when you recognize actual achievements in the reduction of injury and illnesses. OSHA has limited resources, and to get meaningful investigations, you must have enough trained and experienced people to go where the people are getting hurt, sick, or killed. Working with a very limited budget in an atmosphere of deregulation spells trouble for the rest of us. Some people call what OSHA does "windshield inspections." I never heard of anyone inspecting from their car windshields, but when inexperienced people conduct inspections with outdated and inadequate standards, it amounts to about the same thing. They are not serving our national interests by just looking at the scaffold, trench, electrical, or whatever and moving on.

YOU CAN'T BLAME PEOPLE FOR ACTING LIKE DOGS IN A DOG-EAT-DOG WORLD

SO MUCH OF THE WORK OF THE SAFETY PROFESSIONAL IS ABOUT changing the culture, and most of us in this business have seen the dog-eat-dog world all too often. The question for the safety professional is about changing the dog-eat-dog world into something more productive and hopeful. Everybody, no matter what their position, wants to feel respected and spend their time at work in a way that is satisfying. There are jobs that are much more rewarding than others, but the safety professional must take the stand that each and every job can be rewarding and satisfying to the person doing it, or they should be moved into another job. I have seen a good personnel department hire the right people for the right jobs, and I have seen them hire any people they can get. This makes a big difference to the entire culture. The dog-eat-dog world is a way of expressing a lack of trust in our society and other people. It sums up the problem faced by the safety professional or the environmentalist dealing with disgruntled and unhappy people. We always thought we would have a life at work being productive and working together for a better future. When this is no longer the case, there is nothing but unhappiness and resentment.

The world is moving on, and the population has grown, and the complications grow, and more of us feel like it is a dog-eat-dog world. When we feel helpless or our efforts are useless or unappreciated, it makes it a lot harder to continue forward. There can be frustration, and people give up on working toward the good things in life. I can identify with the feeling of a dog-eat-dog world with my giving up on separating my bottles and cans in my garbage for the environment. At first, I was totally on board and felt good about rummaging around in my garbage. After a while, I started to have doubt that it was doing any good. I saw some stories where they were just dumping the garbage I was separating in the landfill. I also know that companies are drilling into our ground and tearing down our trees and mountains. It became so discouraging that I stopped rummaging in garbage. I didn't want to feel that way. I wanted to feel like I was a part of a massive effort that was going to save our planet. It seemed like it was a lie, and it wasn't going to work. I felt a little foolish.

Safety and health professionals have some of these same issues when people feel alone, unrewarded, and powerless in the world or at work. We need to encourage and reward participation! I find encouragement in the fact that preventing accidents and injuries is a money-saving activity that can be shared with everyone who participates in the safety and health effort. I saw a commercial the other day that some company owner got so much money on cash back from his credit card that he bought all his workers health insurance. This is how it needs to work, and this is how it can work, no matter how much chaos there is in the world. If the safety professional can show profits from safety and health, there will always be an interest in this work.

SAFETY SCIENCE

THERE IS SO MUCH TO BE LEARNED IN SAFETY SCIENCE. WE need to know more about the effect that caring for people has on productivity, success, happiness, satisfaction, reduction of injuries and illnesses, and associated cost savings. We can study this issue and document the benefits of a positive, productive culture on the humans associated with the work. Even without studies, I think we all know how we feel when people care about us, want the best for us, and encourage us to be our better selves.

When someone is unhappy, bored, or hurt or killed, the safety professional uses whatever techniques they can to determine the cause. We use deductive reasoning, fault-tree analysis, statistical methodology, and whatever evidence there is to prove that a cause-and-effect relationship exists between the danger and the injury or death. The problem is the cause of that injury might be the lack of a culture of caring and failure to implement safety programs. We can find a cause of a fire, of an electrocution, or a machine injury. We can fix the wire and guard the machine, but we might never get to the fact that the culture in that facility was such that somebody allowed materials to collect that could catch fire or allowed an uninsulated electrical wire or failed to replace a guard for a machine. We can get into a lot of trouble just fixing the hazards. Too often, I see the guard replaced or the wire insulated, and nobody is looking at the underlying cause.

The question is often, why do these hazards exist? If nobody is running a safety and health program, the question comes down to the culture and environment where such behavior is acceptable. There are two important aspects to the science of safety. One aspect is the prevention of an injury before it occurs, and the other is the determination of the cause of the injury or illness after it happens. In both cases, the safetyman or safety woman must decide, and the solution must ensure that these behaviors that occurred in the past will no longer be acceptable in the future. In other words, we must hold people accountable. I am always looking for the name of the person responsible for guarding or electrical. The safety science in most of these cases is about changing the culture behavior and holding people accountable. There is nothing worse for the safety and health profession than somebody saying the answer to a problem is just fixing the hazard and not doing anything about the behavior or culture that caused that behavior to exist in the first place. If a worker falls from a roof, there is the possibility that there was no fall protection, and there is the possibility there was fall protection but that the worker wasn't using it. If there was no fall protection, it is a clear violation of standards, customs, and practices. If there was fall protection and the worker chose not to use it, it brings up the question of culture. The safety professional must find out what is wrong with a culture that doesn't provide safety and health protection and a culture where safety and health protection is provided but not utilized. Just putting up a guardrail or yelling at a worker will not solve this problem.

THE SAFETY PROFESSIONAL IS ON A SEARCH FOR THE TRUTH

In investigation work, there is an old saying that there are different versions of an event, and then there is the truth. The work of the safety professional is a search for the truth. The truth comes in many forms, including verification of the evidence, the facts, the standards, and the expert opinions. The search for the truth about a serious safety issue or health problem is important for survival. Without the truth, the problem or situation that created the injury or illness cannot be prevented. Finding the actual basis of facts of the event is not easy, and finding the truth is even harder. This is most important in preventing the insanity of repeating the same incidents again and again. The dog bit the child. Do we get rid of the dog? Do we separate the dog and the child? Is it possible to train the dog not to bite? Should we get another breed of dog? Can we develop guidelines for dog ownership to prevent dog bites? The concerned safety professional or investigator suggests checking the breed of the dog, testing the dog's temperament, thinking of situations that might cause the dog to act out, and making certain the child doesn't tease the dog.

The point is that solutions to problems are not always simple and require experience, knowledge, and thoughtfulness. You can't just go out and get a dog and not consider the possible consequences. The most important thing is to find the truth to prevent another loss down the road. The safety professional could have another career just going back and checking on conditions years later after an injury or death has occurred and seeing if changes were made to prevent a reoccurrence. I had a case where there was an airboat racing on a river in the Midwest. This was the kind of airboat you see in the Florida Everglades. I had no idea that anybody would be racing these airboats, but I soon found out that they will literally let anyone race these airboats. During the airboat race, an inexperienced driver ran off course and aground. Her airboat hit a family watching the race from the riverbank. The injuries were terrible and devastating for this family, and I was surprised to find out that nobody had planned the race to assure that drivers were qualified. I could see that this was a good case and an opportunity to change an industry that was operating with no organization and planning for safety. Some research allowed me to find that there were some standards for planning for safety at this type of event from the coast guard. With the help of these standards, I was able to assist this injured family in receiving a settlement from the event planners, and more importantly, to get them to think more about protecting their fans. Planning these events became more important, as these races were occurring all over the country, and other races were being planned and organized with no precautions to prevent a similar incident.

In another case, a laborer fell from a racking system where he wasn't provided fall protection, and he was not tied off to any proper anchorage. I read reams of testimony about the lanyard and body harness and whether he should or could tie off to the structure. I was frustrated because nobody could see that the problem was they were treating this laborer, who fell to the concrete floor and had a closed-head injury, like he was an engineer. They were taking a worker with very little education from another country and presuming he knew about the stability of the structure, anchorage, and the capacity of the various elements of the racking system. He didn't know about fall distance or swing radius, and he didn't know—and apparently nobody else (except we safety people) knew—that the components of the scaffolding at a ninety-degree angle would cut the lanyard, even if he was using it for fall protection. I spent three hours trying to teach lawyers that sophisticated engineers must devise a fall protection system, plans, and programs and that these laborers are in no position to know or understand how to protect themselves from deadly falls. These laborers can't do engineering work or calculations. It is the responsibility of other people or parties to make these lifesaving efforts. It would be like a lawyer being expected to know about every type of medical condition.

Both lawyers and companies need to hire experts to explain complicated and specialized concepts and ideas for the protection of the people who are doing the work. The experts can cost money, but my experience tells me that they will save more in the long run. Experts can be quite specialized. One engineer might know how to build a building but might not know anything about aircraft dynamics or motion dynamics of a boat on the water. Lawyers and employers might understand the need for engineering experts, but for many companies, spending money to engineer a safety system for a racking system seems too expensive. They believe that installing a racking system means that day laborers can do the work, and their job is done. But the job isn't done at all if they don't provide the proper equipment and instructions for a determination of where it is safe to anchor their ropes and lanyards. This should all be a part of the plan, and owners, manufacturers, and installers must include this in their bids. The way I explained it was that car manufacturers are responsible for the seat belts and airbags.

Another example I used was when you take your family to an amusement park to ride a roller coaster, that coaster is going to go upside down, left and right with sharp turns, and down a steep embankment, but the manufactures install and devise a fall protection system that will protect the riders under all conditions. It is my desire that the reader understand that these laborers are not sophisticated engineers, and they depend on management to hire people who know how to protect the lives and the families of these workers. It also makes good economic sense to hire these experts because, in the end, they are less expensive than the litigation. This is not to mention the poor laborer who hit his head on the concrete and is severely brain injured but still living six years after the event. It is terribly important that we search for the truth in these matters and demand changes in the practices and attitudes about worker safety. These laborers are experts in hooking together the parts of the racking systems, and they might have been doing this work for many years without an incident, but this does not make them professional engineers.

THE WORKING WORLD IS A DANGEROUS PLACE

Most injuries happen in or near the home. That is where we usually spend the most time. There are plenty of serious injuries that happen at work. Some people spend almost as much time at work as at home. When working at OSHA and for my clients, I saw terrible danger at work, and I went to the site of the injury or death. I have seen workers in a trench with no trench box to hold up the forces of the earth. I taught students about soil mechanics and how unpredictable soil can be, especially when you dig deep into the ground. There are different types of soil and objects, called surface encumbrances, adjacent to the trench that cause danger. The most important factor is the effects of water, which will turn seemingly strong soil into mud. When the soil is excavated and not protected with a shield, a cubic foot of dirt weighing over a thousand pounds will crush and kill a worker. Even if the worker is just trapped within the trench, when they breathe in with their lungs, the void created by the chest compression will be filled with soil, and the worker will not be able to expand their lungs to get more air. They will suffocate to death.

Every time I see someone up in the air with no fall protection, I know they can fall. I have seen them fall, and I have seen them prevented from hitting the ground with lanyards, harnesses, and netting. I had one client who told me he thought I had extrasensory perception (ESP) because I predicted that a wall would fall and that a crane would fail, and both happened after my inspection. This was not ESP at all; this was a dilapidated wall and an old rusty crane. I was just doing my job to try to make them concerned about the hazards that I could see with my own eyes. There are hazards I could see with my fresh eyes and hazards they could not see because they were not looking for hazards. At least ten times, I have told companies that they will have a forklift injury, and sometime later, they called me up and told me about the injury. I could see they did not have the forklift operations separated from the pedestrian walkways. Whenever I see that situation, I know something can happen. I want to stand at the location until they set up the barricades and paint the lines, but all I can do is tell them that what I have seen is an imminent danger, and they need to shut it down until it is corrected.

I invite you to go to any business and tell them that they should shut it down until they make it safe. I have learned this about human behavior, regarding demanding the correction of safety issues: (1) they rarely listen, and (2) they never shut their work down until *after* the accident of injury. I tell them while I am standing right there that I personally, as a safetyman, have an obligation to make sure the corrections are made. I will even put it in writing. I write it out that I have told XYZ Company about this hazard, and they must make the corrections. My brain can't tolerate unnecessary injuries or illness, and it is hard to accept the fact that I am no longer with OSHA. I can't force my clients or other people to do the right thing. I do everything possible to stop the situation and get something done. I have even called OSHA in a situation of imminent danger. It can be a terrible position. I don't own or control the company, and I am standing there, screaming and pleading for them to take some action, and they just don't do it. I am determined to be a safetyman, and I work at it every day. I just don't have the authority to get the correction done at places that are under the control of other people.

When I have worked under contracts for large corporations, I have asked them for the authority to stop the work. On several occasions, this has been a deal breaker. I have found it important to try to negotiate the correction of hazards before I take on a job. I have never stopped learning how to negotiate with people to get a good result. Over the years, I have learned ways to try to get things fixed before somebody gets hurt. Sometimes calling OSHA is the only way. A safety professional goes to places where things are made or built and sees different processes, and with those processes, they can see the danger and potential danger. We see things that most people never even think of as dangerous to people or the community. At a refinery, there are large hydrogen sulfide and hydrochloric acid tanks that most people don't even notice. I know too much about those tanks. A leak or a spill will kill people over a large geographical area, depending on which direction the wind is blowing. I know the importance of the windsock on the tower. I see homes and families near the plant. There are documented incidents where hazardous materials are released in a neighborhood near the train tracks. If you live near the tracks of freight trains or near the chemical plant or refinery, you should be concerned and prepared. I have noticed people act like nothing is going to happen, even when it is obvious that something terrible could happen. I have seen these large propane storage tanks hidden in a yard or on an empty lot. Those tanks could blow and destroy homes and families in the area. People don't understand the magnitude of these explosions. People see a tank and think nothing about it at all. I see these potential explosions.

THE LIVING WORLD IS A DANGEROUS PLACE TOO

There is the weather, the oceans, the lakes, the rivers, the traffic, the animals, and travel by airplane, railroad, buses, and trains. When you take the time to think about it, the world is such a dangerous place, and there is proof of the danger in the news each day. Imagine all the incidents that we never hear about. If we did not have safety professionals investigating, reporting, writing rules, and enforcing rules and regulations, it could be much worse than it is now. There are elevators, escalators, and amusement parks. Too much heat, too much cold, sun damage, bacteria, disease, and I can't name all the dangers in this wonderful world. Without government and private agencies to protect our food, water, and buildings, we would be in real trouble. There are the amazing codes from the National Fire Protective Agency (NFPA), National Electrical Codes (NEC), consumer products regulations from the Consumer Product Protection Agency, and hundreds of other agencies and fine people who write and enforce rules and regulation that protect us and our families and friends. Without these people and these rules, we might all be gone by now.

When I first became a safety consultant and safety expert, I was focused on the working world because I was with OSHA. There was plenty to do because when I started in the early 1970s, it was a cowboy world where safety and health considerations were just emerging. They were sending me wherever people were getting killed, injured, or sick, and I had a big job just protecting myself. Below are pictures of me entering a confined space and getting ready to enter the steel mill.

As time went on and because I am a safety expert, I began to realize that the danger and the hazards that were inside the factories and on construction sites were also outside those areas. I started looking into the hazards that affected people who were not working in a facility. I saw safety and health hazards to which we are all exposed. There are the chemicals we buy in the stores, the vehicles that are on the road, and the products we use every day. The safety and health dangers in the world outside the workplace were just as fertile as inside the workplace and maybe even scarier because of few or no regulations. Inside the factory or on the construction site, at least there were OSHA standards to enforce. Outside, there was no enforcement of regulations, fines, or punishments for violations. I just listened to a recording of a telephone call to 911, where a mom found her son drowned and with no pulse. There is no horror movie or story than can be worse than listening to that call. This telephone call is now permanently embedded in my heart and soul. It reminds me of why I do this work.

Safety and health must be represented everywhere because this world is so dangerous. Horrible life-changing events happen to people every second of the day. I recently got a small cut while cutting an avocado when it seemed to slip right off my knife, partly because of its round nature and hard pit. That knife blade headed toward my left hand. I realized that I should be doing something to protect myself when I am cutting avocados. I went to the internet and found out that thousands of people get severe injuries cutting avocados. The only people who know how common this injury occurs are the medical professionals in the emergency rooms. Just another case where safety needs to be part of our everyday life. Snow shoveling is deadly, as is ice, and then there are holes, tripping, electrical hazards, tools, and so many other ways to have a serious injury. It's a dangerous world, and hardly anyone is working to change it. There must be improved safety reporting on the internet or on television about what is going on in the safety world. If car after car is damaged in a gaping pothole or if electrical lines are down in the neighborhood, we need to be informed. Riding the commuter railroad in my area has become unpredictable because of frequent suicides. I need to know immediately if the train is shut down. I learn of new and different dangers in the safety and health publications I read every week and at the safety seminars and meetings I attend, but the average person gets very few warnings unless numerous people get killed.

I heard recently that a large furniture company had to pull their dressers off the market because kids were getting hurt when they fell over, and several children had died. Apparently, there were few or no instructions or warnings about the necessity of securing these dressers to the wall. Whenever I see a dresser or a file cabinet with the top drawer open, I think it will fall over. Sometimes they do fall over, and the weight can be significant. I have had a few cases where the weight of a file cabinet, a storage cabinet, or a racking system came down and maimed or killed people.

There are hazards each of us faces every day. We become accustomed to and accommodated to the danger. Each of us may face violence, carelessness, fighting, hostility, rage, and a whole range of human emotions, a sort of game of thrones in a dog-eat-dog world. Nobody is taking the time to think about or plan for the danger they face every time they leave the house—and even within the house. The safety experts' perception of the world is that it is a snake pit, and the snakes like it that way. If you are the strongest snake in the pit, then the wars are productive and fun. Having the world be a dangerous and selfish place is the biggest problem we have.

The more danger and the more survival of the fittest we have, the more need for the safety expert. Sometimes, the poor and helpless are exploited, and they are not able to stand up to the power and authority. There are also many social pressures on individuals not to complain or make any waves. Nobody like complainers. This makes it more important to establish rules for protection and preparedness and being proactive.

Dangers faced by the poor and helpless and all of us seem to get greater as our society becomes more complex and as we age. We have more people, more congestion, and more frustration and anger. People have different beliefs and ideologies and they can be erratic and unpredictable. All of us have seen people behind the wheel angry or driving in an uncontrolled manner. All we need to do is be in the wrong place at the wrong time. At this point, many of us don't feel safe driving through the green light because some distracted or impaired driver might not notice the light is red.

We don't all walk the same, and we don't always act the way

someone else might expect us to act. Many of these social problems might never be solved. To the safety professional, every human life deserves protection. The alternative world of the safety professional is a world of safety and security for everyone. A world where people look out for one another and where the authority of the world writes, communicates, monitors, and enforces standards, customs, and practices to ensure compliance with basic human rights. We believe in life, liberty, and the pursuit of happiness, and to achieve these things, we need to have a safer and more healthful world. Sometimes it seems we are going the other way, taking more chances and not fully understanding the consequences. We try to reduce the greed and competition for financial gain and contain our lust for danger.

The safety professional believes that many of the dangers we accept in our world come from a place in the past when we were not sophisticated enough to develop strategies to prevent catastrophes. We believe by sharing our experience and information about the danger, we can change the world. It is disturbing when those who have been fortunate enough to benefit from our prosperous society do not take the action necessary to protect the people and the earth. Those who are fortunate enough to benefit from the natural resources of this planet should be responsible for protecting the people and the planet from the danger that results from their actions. The earth and the people of the earth need protection. Can you imagine (John Lennon) how different the world would be if we were all stewards of the earth and of the safety and health of the people?

MULTIEMPLOYER WORK SITES

The concept of the multiemployer work site is probably the most important concept in safety over the last thirty years. When most people hear the word "employer," they think only of the person who is the boss or is paying for the work. On most projects today, in construction and in industrial facilities, there are many different employers interacting with one another or with the production. There are specialists of all kinds. There are cleanup crews and vendors, and each one of those employers that is not the employer of the employee can bring hazards and danger into the workplace that cause injury or illness. These other employers must be responsible for their own safety and health programs and coordinating with the other employers at the work site or facility. Originally, it was thought the only party that could protect workers' safety and health was the employer of the employee, and most of the time, it was this employer that was held responsible for the safety and health program and the safety and health of their workers. The problem was as companies and workplaces became more complex, there were more employers working at the same location with more complicated interactions. Any of the employers could create hazards that would expose other employers' workers to danger. When more than one employer is working at the same location and interacting with other contractors, it becomes increasingly important for a single employer—the controlling employer—to arrange, stage, and coordinate the work activities for the entire plant or jobsite. These controlling employers, often called general contractors, prime contractors, and construction managers, have the responsibility of coordinating and staging the work so that the safety and health effort works for everybody at the site. OSHA calls these controlling employers, and I call them the conductors of the symphony of construction or industry. There are also other employers who are not controlling employers and may be conducting activities, affecting employees of another employer or the workplace or jobsite. Those employers have to be held accountable too.

In OSHA, there are four different types of employers: (1) controlling employer, (2) exposing employer, (3) creating employer, and (4) correcting employer. The controlling employer is the general contractor, prime contractor, owner's representative, or plant manager who coordinates and stages all the activities for safety. The creating employer is any employer who creates hazards for their employees or somebody else's employees. The exposing employer is usually the employer of the employee, and the correcting employers are employers who have a responsibility to act on hazards and fail to take an expected action. If you think about this concept, you can see that without it, there would be no real possibility of significant safety and health effort. It is a shame it has taken so long for this simple idea to be adopted by industries and construction. To this day, I run into people, sometimes managers and other safety people, who believe that each employer is responsible only for their own workers. The safety and health program can never work in this way. There must be somebody who is planning for fire protection for the whole plant or construction site. Someone must plan for emergency evacuation, porta-potties, housekeeping, fall protection, and many other activities that need to be coordinated on the project. If an organization comes to the conclusion that it is every contractor for themselves, and there is no planning and coordination between the parties for overall safety, I think you can see it would be a chaotic situation, and accidents and injuries would be inevitable. It would be like an intersection with no traffic control.

SEPARATION AND ISOLATION

Separation is one the best engineering tools we have to improve safety in a world of moving vehicles, equipment, and materials. Things can go wrong, and separation is often the difference between life and death. The concrete or jersey barrier is one of the most important developments to come along in safety. This modular concrete barrier separates traffic and protects buildings, pedestrians, bicyclists, and many people in the age of potential terrorism. It is called a jersey barrier because it was developed at a college in New Jersey, and it weighs approximately six hundred pounds per foot. It is almost always ten feet long, but it can be twenty feet or thirty feet long. Precast concrete is the most amazing forming building material, and here in the form of concrete barriers, it is one of the least expensive methods for separation. Many other materials are used for separation, including plastic, sometimes filled with water or sand, and many kinds of balusters, some that open and close mechanically.

There are too many vehicles in cities today, and as the danger from terrorists keeps increasing, it is unthinkable not to contemplate separation in almost every location. There have been attacks where vehicles are used to intentionally do damage to pedestrians, and separation has become an essential consideration for safety and security. There are so many kinds of vehicles, including semi tractors, trucks, vehicles, motorcycles, bicycles, and scooters. Evaluating the interaction between these vehicles and pedestrians is critical for both safety and security. Separation and preparedness are the primary tools to prevent disaster. Many people report close calls or near misses, where there is not adequate separation. Pedestrians and people operating vehicles and equipment will make mistakes. They are in a hurry, distracted, and fearless. When there is no clear separation between the vehicles and the pedestrians, there will be accidents and injuries.

Safety engineers have been working on solutions. In many cities, you can see ingenious methods of keeping the people away from the vehicular traffic and even separating the bike riders from the vehicles and the people. Unfortunately, most of the world is behind in dealing with the separation problem. In many communities, there are not even sidewalks. We have no sidewalks in my community. If we dare to walk near the street, we are taking our lives in our hands despite the thirty-mile-per-hour signs. The concept of separation needs to be adopted by every community for the protection of its people. Isolation is like separation, but it offers even more protection. When we say that a hazard is isolated from human contact, it implies that it is beyond separation. With separation, someone could climb over a barrier, but with isolation, it is not possible to get into the hazard area. We use the term isolation with energy and especially with electrical energy. We say with lockout and tag-out standards, the sources of electrical energy are isolated, meaning they are fully identified and protected. The concept of isolation can be applied to many other aspects of safety and health. It would be a good idea to isolate the railroad tracks from pedestrian and vehicle traffic. There is more isolation from highway traffic than for railroad traffic.

SAFETY AND SECURITY

Security is safety. People think that security is security guards. Security is beyond security guards. Security is a plan for safety that identifies hazards and dangers from the outside that must be evaluated, and action is taken to minimize the danger. Security safety has become the subject within the school system, where there have been significant breaches. There have been discussions of armed security guards, metal detectors, and even armed teachers. It is not just the schools that need these kinds of protections. Every place where humans congregate needs an evaluation. Safety professionals must develop security systems with their safety programs in today's world.

There are experts who specialize in security systems, but I would want my security experts to also be safety experts and well trained in human behavior. There are bad people with bad intentions in this world, and we are going to have to deal with them the best we can.

Security and safety are the same thing. A safety audit should never be conducted without consideration of security issues. There must be separation barriers and security fences around the property, as well as cameras at every establishment. Most public places and private companies should employ the use of metal detectors. Establishments where people congregate cannot afford to have even one disgruntled person or worker get on the property with a weapon. Sports and concert venues seem to have the most elaborate security systems, but there can be serious defects in even some of the best venues.

Information and technology are changing at a rapid pace, and those security systems need to be evaluated and upgraded on a regular basis. Even the best security systems, such as those at our airports, fail on occasion. A complete safety review must include security. The security of the perimeter should be evaluated, and the security of the personnel is a separate concern. Background checks are essential. OSHA (and most safety programs) has a section to prevent violence in the workplace, and these standards and programs can be used to formulate a program that is proactive and anticipates the possibility of someone trying to destroy the property and hurt the people.

I had a company that manufactures precast concrete that failed to use fences and cameras to secure their property. Despite knowing that materials were being stolen from the property for many years, they continued without a security fence or cameras. A couple of young children came on the property to steal some metal parts. Some of the metal was holding the precast concrete in a vertical position. When they pulled the steel support from the concrete panel, it fell and killed one of the kids. The questions became not about the trespassing or even stealing; it was about a reasonable security system. It was my opinion that a facility with these heavy, dangerous, precast panels must have security, including separation and cameras. I think every plant needs it.

ENJOYING THE WORK

It is difficult to be a safetyman because people don't like to be second-guessed or criticized. Despite these problems, I love being a safetyman. I get to make a living and do something worthwhile in this world. Enjoying my work and caring about people makes safety work different from other jobs. I get more satisfaction out of my work than most people. Working hard fighting for safety stimulates my brain and keeps my mind away from thinking negative thoughts. I like any job where I get to interact with people. Too many people just want the workday to end so they can relax and enjoy their lives. Many people in my life are focused on their retirement when they no longer need to work. I was surprised how often the people who worked with me at OSHA talked about retirement. Many of them could recite the date of their last day of work. I am not sure why I enjoy working so much, but I think it is the work of being a safetyman. I tell people that working is so much better than sitting home, watching television. Lately, I have been getting joy out of writing and playing my guitar, but both seem like work. Writing a book and playing guitar are difficult but so rewarding. For me, being in safety is about caring for other people, but it's also about enjoying my work and my life at the same time. Don't get me wrong. I like to ski, and I like to be with friends and go out to dinner and see a great play or a movie, but believe it or not, I enjoy my work as much as any of those activities. There are a few other people I have met in my life who enjoy their work every day but not that many. I would recommend it to anyone. It has made my life happier and more fulfilling. I have a feeling that I am doing something good and being productive at the same time.

HAVING GOOD SAFETY MANAGEMENT

COMPETENT PEOPLE

IN THE SAFETY BUSINESS, THERE IS A TERM CALLED THE COMpetent person. This is one of the most important concepts in safety, and it should be adapted to regular life. This person has the training, experience, and authority to take appropriate action to protect the safety and health of humans. This concept puts the responsibility squarely on management for safety in organizing and planning work activities. The same concept can be used by adults to protect their families. Somebody needs to know what they are doing before they go out and buy a chain saw. People and companies want to rely on everyone to protect their own safety, but this idea doesn't work because people without experience don't anticipate all the potential dangers in using the chain saw and many other activities. The individual person using the tool or doing the activity may not see all the potential problems or the big picture.

I never had any idea when I cut avocados that the knife would hit the pit and head for my left hand. If I had done more planning, I could have gone to the internet and found out how many people had been injured cutting avocados. This is the same problem when using a tool or conducting a task where one person has no control over what other people might do in the area or assisting with the task. One person, worker, or contractor might leave a hole uncovered or remove a guardrail, unknown to another unsuspecting person. The competent person is the person in charge. It might be a manager or foreman whose job is to be watching and making certain that things don't go wrong. Trusting another person to look after the safety and health of another person is not a program of protection. We need individuals who are responsible and held accountable for safety.

WORKING FOREMAN

There is a concept in the trades called a "working foreman," which is contrary to the concepts of safety and the definition of the competent person. When a working foreman is concentrating on their work, they are not spending the time that is necessary to see the entire picture. When a competent person is the one who is protecting the person working on the ladder or watching for the power lines, and that person walks away to conduct some other task, bad things happen. I do not believe in any working foreman having safety and health authority. A person assigned to protective responsibilities already has enough to do. I have seen the person who is supposed to be a safety monitor, watching people on a roof so they don't get close to the edge, turn away for just a few seconds, and the results have been terrible. Just tell the safetyman that you have working foreman with safety and health responsibilities, and I will know everything I need to know about your safety and health effort.

HAVING AUTHORITY TO ACT

Authority is one of the biggest issues in safety. Any competent or qualified person must have the authority to stop the work or take action to protect the people from an unsafe condition. Having authority and using it are two different things. Nobody wants to shut down an operation. When you shut down an operation, you have the attention of everyone at the plant or the construction site. The spotlight is not a good place to be, feeling the anger of everyone who is trying to make a living. People, even competent and qualified people, fear the responsibility of taking action that will interfere with the big picture, namely production and profits. They might think that they are doing the right thing by shutting the operation down, but then, if nothing happens, they will be criticized for shutting down production or a machine.

I believe this is what happened at Deep Horizon with the explosion and oil spill, where people in authority failed to shut down the producing oil well. This has also been a problem in other situations where safety and health warnings have been overlooked. It is an especially brave man or woman who will stand up to the social and production pressures to stop an operation or activity that is producing income. How many people do you know that have guts like that? We need to train people to take this kind of responsibility without fear of the social and production pressures. If they shut down a process, a machine, or a line, and they were wrong, the price is much smaller to the company and society than a massive oil leak into the ocean or an explosion. Such action might prevent a severe injury or death, and they will not get any credit for preventing something that never happened because of their actions.

Today, I learned that a crane fell in downtown Seattle. I will not be surprised to find out that at least one person knew something was wrong but feared to speak up until it was too late. The Hard Rock fell in New Orleans, and when they were questioned, numerous workers gave statements that they could see the bending and stress in the structure before the failure. I have learned how difficult it is to be a person in authority and concentrate on important issues, as well as the difficulty of a safetyman or a competent person trying to keep their eye on the ball. It takes so much courage, and we don't train our people enough on the value of being courageous. We must be better at encouraging our leaders to stand up for safety and health. I have a vision of a safety training program for the development of leadership where they get some experience being brave in the face of production pressures, social pressures, and higher authority. There is a program from DuPont called STOP that teaches each worker that they have the authority to stop the work, and they will not be punished or humiliated. This program has been successful and must be expanded until every worker has the courage to take action that is needed.

WHAT IS A GOOD SAFETY MANAGER?

When it comes to management, it's the survival of the fittest. But the question of why management people are the fittest is another matter. Management people need not only be production people but also people who care for other people. Too often, I find management people are production people only, and they are cracking the whip for higher production and more profits. The kind of managers we need today are the ones who lead subordinates in a way that makes them proud of their work and encourages them to work productively and safely. Safety and health people make the best managers because they have demonstrated an interest in people. If you can solve safety problems, you can probably solve any problem. Solving a safety problem requires attention to detail, analytical skills, and the ability to persuade other people to adapt to change.

Identifying safety people is one of the things we do for our clients. We interview workers and look for an interest in improving culture and conditions. When we find these people and think they are good candidates for management positions, we note it for our clients. Many of our clients have told us that this effort was one of the most helpful in upgrading their safety programs and their corporate culture. Word gets out that caring about conditions is a path to more pay. The safety people need to be on their way to the board of directors. Losses in the form of injuries and illnesses will put a company out of business in a flash, and savings from the prevention of accidents and illnesses is rarely credited to the safety managers who accomplish life-saving goals and objectives.

Traditional managers were totally production oriented, and there hasn't been enough change from the old ways of doing things. Some of the people who hold the power don't want to give it up to some safety expert. One day, the safetyman and safety woman will get the power they deserve by being not just production people but also protection people.

PRODUCTION/PROTECTION PEOPLE

Most of the time, the safety director is midlevel with no actual authority over production. This does not mean things won't change, but from my experience, there will be more safetymen and women moving up to higher and higher positions as their value increases the bottom line. I think things will get better. The safetyman sees more regulations and more interest in safety in the workplace and in the world.

The production people currently in charge of operations fear losing control. Production people need to understand that being safety people will only enhance their careers. It is not possible today to be in a management position and not be in tune with safety and health concerns. Management people with a poor safety attitude are nothing but trouble for modern corporations and construction sites. Get them out of the way, or eventually, it will result in a costly injury or illness. Those people not supporting safety goals are a serious obstacle to the success of my clients. Successful organizations need to have production and make money to pay for good safety managers and good safety cultures. Having a person in a leadership position with a bad attitude about safety is the same as having a serious safety hazard.

I open communications with clients with the production/protection concept. Production and protection go hand in hand, and the people who oversee operations need to know it from the beginning. I often see that the safety work is not fully integrated into the production operations. By creating titles and positions that are both production and protection, we get the right message across. We want production, but protection runs side by side.

HIRING PROFESSIONALS

People have special skills and spend their lives and careers learning how to do their work in a professional manner. Too many people think they can do everything themselves. Few people have a knack for working with tools, but amateurs use tools every day. Without experience, it is easy to get a drill bit in the eye or wreck your knuckles with the sprung jaws of a pipe wrench. It takes years of experience to learn how to do the work in the proper way, including the techniques and tricks to get good results. To get it right, people need to do the same task over and over until they learn how to do it right. It seems like everyone would know this and hire a person with some experience to mow the lawn, use a chain saw, or fix the plumbing. Unfortunately, there is the element of saving money, and the ego becomes involved when we decide to do it ourselves. Many people must think they can just pick up a violin and play Beethoven.

Too many people think they can handle a dead tree by going to Home Depot and buying a chain saw. One mistake or distraction with the chain saw, and a life can be changed forever. How many people must be maimed by chain saws before the word gets out? It is cheaper to hire someone trained and experienced in the use of the chain saw and who has all the proper protective equipment. It's probably cheaper than the chain saw itself.

We all know people who should never be allowed to use a chain saw or a forklift or any dangerous piece of equipment, but the rule in our world is that we all have the freedom to do whatever we want. I am all for freedom and believe they should be able to do whatever they want. But they should be informed of the alternatives and the consequences before they act. No matter how silly or dangerous their action may be, as a safetyman, I would recommend hiring a professional. People need more respect for the skills and training that another person might have to get the job done in a safe and productive manner.

There is a new trend of telling lies on job applications, which really gets to the problem of not having qualified people. On the late-night shows, the actors always say that to get acting jobs, they must lie about whether they can sing or dance to get the job, and once they get the job, they just do their best. The trouble with the rest of the world is that we are not acting. We are cutting down trees or building houses, and when we lie about our skills and abilities, it ends up as a tragedy.

Twenty years ago, in the basement of my house, I decided that the area between the utility room and recreation room would look good separated by a small glass block wall. I had experience watching other professional craftsmen do work, including glass blocks and concrete, and this project seemed to be a project right in my sweet spot. I failed to get any professional assistance or advice and bought the glass block and the grout and went to work. After about five hours of work and overnight drying, I could see my mistakes. This wall was not straight and not secure. It was put together well, but it was not supported on the sides by the adjacent walls. After it dried overnight, I pushed a little on the wall, and it fell over and broke into many pieces. It would have been much cheaper and safer if I had just hired somebody who knew what they were doing. It was a good lesson for me and for all of you who want to do it yourself. Get people to do the work who know what they are doing. I realized that the particulars of putting up that glass block wall are more than just putting the glass blocks together. I had no experience supporting it with the adjacent wall. After talking to some people who know how to do this job, I learned that it is not easy to fix.

WORKMANSHIP

A man was picking up debris from the veneer of a restaurant they were renovating. While he was bent over picking up debris, a capstone from the top of the building came down and crushed and killed him. There are a lot of questions about how this could happen and why the capstone wasn't properly anchored. It was not anchored for forty years. It was sitting on a brick veneer that is not considered to be a structural member. The level of incompetence that resulted in the death of this poor man was astounding. It was a lack of professional workmanship. When they built this building, it would have been easy to anchor these thousand-pound capstones. Why place them on veneer that is not a structural member? When it is built sloppily and you go to remove the veneer with a chipping hammer, the lack of stability and vibration will cause it to fall.

There were also questions about the demolition of the building and the lack of a demolition survey. Where were the engineers to make certain that there was stability when these laborers were working under the danger? The mortaring of the bricks that were holding the capstones was substandard. For more than forty years, the unprofessional work was a time bomb, waiting to explode on some unsuspecting person. Despite the time bomb, the general contractor and the demolition contractor must do surveys and evaluations of the structure to make sure it is stable before conducting any construction or demolition activity on the structure. How could they be chipping on the veneer of this building without examining the building thoroughly? It seems impossible, but it happened, and a man was fatally injured. I wonder how many other structural defects exist in these older buildings and how many times the general contractors and engineers don't examine the structure completely before the work is started.

DANGER IN BEING THE SAFETYMAN

THE SAFETY PROFESSIONAL IS EXPOSED TO SO MANY HAZARDS! It is a dangerous job. I once led a group of inspectors at OSHA to inspect fireworks factories after several had blown up. Investigating the fireworks factory or warehouse *after* it has blown up is extremely dangerous because there is still smoldering heat and fire and live fireworks buried under the debris. It was like working in a minefield. One time, when investigating an electrocution fatality at a mushroom factory, we found the ground was energized. If we had stepped in the wrong place, we could have become another fatality.

There are few jobs or occupations where a person will face a greater variety of danger than being a safety professional. A safety professional seeks out and goes to the places with the greatest danger. We might be investigating an accident or a fatality where someone has already been pulled into a machine or electrocuted. We might be at a previously unvisited location, looking in pits, going down stairs and under machines and equipment, not really knowing what to expect. We are constantly exposed to dangerous moving parts, electrical hazards, and contaminants. We try to never to go anywhere that even looks remotely like a confined space unless there is proof that there is enough oxygen to breathe. We need that oxygen to stay alive. We frequently get exposed to chemicals, dusts, and asbestos. At least we can make certain that there is enough air. We are fortunate to be alive, telling these stories after so many years of exposure to danger, and there have been close calls.

One time, I had just stepped away from my car to inspect a smelting operation, and I was still holding the door handle when my car was flattened by a semitrailer. The worst danger in my career as a safetyman has been from irate people. One time, I went to investigate conditions at a scrapyard, and the owner started shooting a rifle over my head. He made it very clear that he was against the federal government and was not going to let any federal agents enter his property. I have also had experiences where the people I was trying to help got angry when I found things that needed to be fixed, especially when it was going to cost them money. I remember going to a machine shop and finding twenty punch presses operating without any guards. I spent three hours writing up all the violations, and the owner was so infuriated and emotional that he gave me the keys to his company and told me it was all over because he could never afford to make the repairs to comply with the law. I believed him, and I felt bad, but I would have felt even worse for the workers and their families when some of them lost a hand under those machines. Looking back, I do think the worst situations involved the thousands of people who looked at me as nothing but trouble and wanted me to go away. If you invite me over to your house, and I find electrical violations, furniture not attached to the walls, no handrails, and some serious danger in the backyard with the kids' castle or trampoline, will you be happy I was there to give you a free inspection?

OUR BODIES ARE SOFT

Our bodies are like butter. I just saw a man at a department store who has the job of disassembling the shelving. He was a strong young man who thinks he is made of steel. I was reminded of the truth of the human body. Most of us act like we are armadillos with a tough armor protecting our soft underbelly. However, the case with humans is that the skin is amazingly soft. It will be cut by a piece of paper; in fact, I was so interested in paper cuts because I get so many of them that I investigated the science of paper cuts. It turns out that a single piece of stiff paper is like a knife when it is in a pad or a pile and has a corner sticking out that is supported by the surrounding paper. It's because it's as sharp as a razor. This would not be such a problem if the human skin wasn't so soft.

The man who was disassembling the shelving lived in a hostile world. I wonder how much training the shelving disassembler had for doing his job and what kind of protective equipment he was using to protect his soft tissues. Did he have cuts on his hands or scars from previous work? Questions of the safetyman. There are many materials and devices to protect our bodies, but do we use them? Gloves, glasses, hard hats, shields, body protection, fire and heat protection, and now they even have suits with heat and air-conditioning. There are so many engineered solutions for the protection of our soft bodies, but how do we get people to use that equipment? Nobody needs to live with the scars of exposure. Too many visits to the emergency rooms. That paper cut might be no big deal until MRSA, antibiotic-resistant bacteria or flesh-eating bacteria, enters the picture, and then it could become a big problem.

THE EFFECTS OF EXPOSURE MAY BE IRREVERSIBLE

Just a sentence but an important concept. Many people don't realize that a single exposure to mold, asbestos, silica fiberglass, and about a million other things might change your life forever. We should be uncomfortable around vacuum cleaners that create clouds of dust. At my health club, they were vacuuming the ventilation ducts, and the locker room was literally filled with dust. I almost had an injury trying to get out of there fast enough, yet there were at least twenty other men in the room who acted like dust wasn't a problem. I stopped one of my friends and told him to watch out, and he just smiled at me, as if to say, "Oh, the poor, crazy safetyman" and went directly toward his locker. First, we have no idea what kind of dust or mold we are dealing with. We have no idea of the material or concentrations, but as a longtime safetyman, I can assure you that when you see clouds of dust, it will have some effect, maybe mostly temporary, but it could be permanent. People act this way only out of ignorance. Maybe if they had seen all the black lung pictures I have seen or if they had met the coal miners carrying oxygen cylinders, they would wise up.

It is amazing the things that I see people ignore. I wish I could think of some words to make you understand how helpless I feel when I see people exposed to deadly chemicals, vapors, or gases, and they just don't seem to care. They don't know what irreversible health effects could mean to their life or family. Irreversible health condition means forever. People who are affected by tragedy gain more respect for the danger after it is too late. I have a friend whose grandson was exposed to chemicals that led to leukemia. That young man may never lead a normal life. They are finding out more and more about the danger of weed killer that causes cancer. I have used it myself. I wear gloves but never a canister respirator. I should have been wearing a canister respirator all along, and I am a safetyman!

INHALATION AND SKIN ABSORPTION

Underrated as a source of cancer and contamination are inhalation and skin absorption. Furniture stripper is especially bad in both areas. Gasoline is bad, lighter fluid, super glue, and so many other things we use every day. The right kind of gloves and a respirator are directed on the label, but even those protections might not be enough to keep you and your family safe. If it gets on your clothes, and you wash those clothes in the family washing machine, you could be exposing the entire family. When you read the label on most of our household materials, you can't help but be fearful of an exposure that could affect your health. The problem is that unless you are a safety professional, you probably don't read the labels or even want to read the labels. This reminds me of all the drug commercials that are on television where they tell you how great the drug is, but in the end, they tell you it will probably kill you. They have one for skin problems that says the scaling, redness, and itching will be gone, but you might die. As a safetyman, I have learned to try to stay away from weed killer, drain cleaner, and other harsh chemicals. I might not get sick, but there is no reason to take a chance.

LOOK UNDER YOUR SINK

I want people to be interested in safety and health, and sometimes they think it is so dry and uninteresting. I am not certain what the right words are to say to gain the reader's interest because this is about your life, your safety and health, and your family. It's about your kids and everybody and everything you love. I know that nobody talks about these things, and I know the reason that nobody talks about the things I work with every day is because if you start down this road, it can be terrifying. Some people call it going down the rabbit hole.

Since I want us to go down this road together, I think we should start with something familiar and terrifying. I already know that you don't want to read all the warnings that come with your prescription medicine, because after you read all those warnings, it is impossible to swallow that stuff. The problem might be worse than you think because it is likely that your doctor didn't read all those warnings either. If doctors read all the small-print labels for the drugs they prescribe, they wouldn't have any time left to see patients. I think it might be useful if we look at the warnings on some common products we have in the home. At random, I picked out a granite and stone cleaner under my kitchen sink. I will not name the manufacturer because there is no reason to pick on one company. This stuff says the following:

> DANGER: EXTREMELY FLAMMABLE CONTENTS UNDER PRESSURE. Do not use near heat, sparks, or flames. Vapors may ignite explosively and cause flash fire. Prevent the buildup of vapors. Do not smoke. Turn off pilot lights, stoves, heaters, and other sources of ignition during use and until all vapors are gone. Do not puncture or incinerate container. Do not store at temperatures above 120 degrees Fahrenheit. Use with adequate ventilation.

Keep Out of Reach of Children, First Aid Treatment. If swallowed, call a poison control center or doctor immediately. Do not induce vomiting. If in eyes, rinse with water for fifteen minutes. If on the skin, rinse well with water. If breathed in, move person into the fresh air.

I must tell you that while I was typing these warnings, the closed container of this stuff started burning my eyes. You might be saying to yourself that you already knew about all this stuff even though you don't read these warnings any more than anyone else. But how about the part about turning off the pilot lights? Who does that? Also, if you blow out the pilot lights, you will be leaking gas.

These problems with chemical agents include 1) people not being able to read the labels; and 2) if you read the label, you will be too afraid to use the product. These manufacturers want us to use their products, so they do everything they can think of to keep the information away from us or make it difficult for us to read the information. A safety professional has got to find a way to inform the public about the danger and provide simple solutions. I know you won't read those labels, and I know if you do read them, you might try to ignore them to the extent that is possible. Nevertheless, these labels are the basis of every chemical health hazard program we have. These labels are the basis of OSHA's Hazard Communication Standard. This is a law that requires a warning to be provided with hazardous products, and to give users of these products information and training about the hazardous nature of the products and the action necessary for the protection of the user or consumer.

The basic protection for our lives and our families is information that nobody reads, and in many cases, people want to avoid reading it if they want to use the product. Can you see what a big problem this is for the safety professional and all of us? Remember, this is just one product I found at my house. I would not be surprised if there were hundreds of products just as dangerous and just as harmful, and I haven't read the labels. I am certain that my pilot lights are too close to those flammable vapors, and I should never use this stuff without special chemical-resistant gloves and an approved respirator. At least I have that stuff.

The problem is that we are not adequately protected from these products and materials, and the protection we have is not useful if we are not going to read it. I think when you buy hazardous material, you should have to prove that you have the proper protective equipment and some training to know how to use it. Of course, this will not protect your kids or your pet unless you have a locked, ventilated cabinet at least twenty feet away from any source of ignition.

Now, I know many people don't want to spend the time and attention addressing these problems, but the fact that you are reading this means you might have some interest in protecting your kids or your dog. You need to take an inventory of the stuff you have in your home and decide what you are going to do with it. No matter what, please do not flush it down the toilet or throw it out in the yard. This stuff will need proper disposal, and it may cost you a few bucks to get rid of it. Get rid of it if you can, but if you think you might need it, you are going to have to read that label and determine the best strategy for where to store it. This is what the safety professional does!

TRICHLOROETHYLENE

There are cancer clusters all over the country where they have found TCE. It's a colorless, volatile liquid produced in huge volumes to make hydrofluorocarbon chemicals, especially refrigerants. It is also used as a degreaser. Short-term exposure to animals gave them nervous system damage, liver, respiratory systems, kidneys, blood immune system, heart, and body weight disorders. In humans, it is associated with autoimmune disease and cancer. It short, it is some really bad stuff that has been declared a human carcinogen. When I was an OSHA inspector, I found this stuff in almost every plant I inspected. When I asked how it was disposed of, which is really an EPA question, nobody wanted to show me disposal records. I have seen people wash their hands in this terrible chemical. How bad is the pollution from this TCE and how many cancers it has caused will probably never be known, but it is a good example of how careless we have been with human life over the years. Too many people are saying that they didn't know how bad this stuff was in the olden days, but the truth is that they knew and didn't care. The first time I smelled this chemical, I knew it was bad.

GERMS: SPEAKING OF CURSES

The terror of small things is a part of nature and the safety world. I think of the painting *The Scream*, also called *The Scream of Nature*. I have a picture of *The Scream* that my wife got me for my guitar-playing area because there is a similar picture in my pick guard on my Martin. Small things are mysterious and sometimes contagious. I can remember my father telling me that he was not afraid of sneezes. He actually would say to me and other people, "Go ahead and sneeze on me. I don't care. I'm not afraid." I think the truth is that he was afraid. Then there are the legions of people who have the theory that the microbes are actually good for you because they keep your immune system working. That makes sense if you are not the one getting the cold, the flu, measles, or Ebola. This is not to mention the peanut allergy.

The safetyman has seen so many things that he has become somewhat of a germophobe. I have seen people become sick and die from exposure to toxic gas, vapors, and particles. The safetyman knows the value of a respirator or even a dust mask when it comes to protection. The safetyman hates close talkers and unnecessary touching. He uses but fears public transportation and public restrooms. The safetyman is careful not to touch doors or handrails with his bare hands. The safetyman knows of people who try to kiss him on the mouth, and he turns his head at the last minute. The safetyman gets a little angry when someone sneezes in the open and not in their armpit or in a tissue. The safetyman has a fear of swimming pools and hot tubs but still uses the swimming pool and the steam room at the health club, knowing he has gotten sick and sinus infections from both since they are breeding grounds for bacteria.

When you think about it too much, humans are just bags of bacteria, bodily fluids, microbes, and germs. History is full of epic diseases that wiped out populations of humans and animals. After a close examination of the situation, it does seem hopeless, but still, we live on and suffer sickness and recover while we are young. It's only when we get old or sick that the danger really begins to reveal itself. The old are much more susceptible to illness and death. It's a wonder that any of us leave the house, but worse, the young people who get sick or have the flu refuse to stay home to protect the rest of us. This is more evidence that we don't look after one another. The only way we can get people to stay home is if doctors or the government put them in quarantine, which is just another word for confinement. Too many people want to get out of the house and make everyone else sick. Besides, if the kids get sick and causes the parents to stay home, they will have to miss work or pay for a sitter. This situation with germs shows the truth of selfish human behavior, which is the bane of safety and health professionals. Humans today can travel all over the world, and in just a few hours, something bad can spread from continent to continent. These concerns are more evidence of the need for the safety and health professional. We need more people who understand human behavior and can prevent the unthinkable.

ERGONOMICS

It wasn't until twenty years ago that I began to realize ergonomics was what I had been studying in graduate school. They were calling it "behavioral cybernetics," or at least that is what Dr. Smith called it. Dr. Smith had a big thing about having the world accommodate the human being rather than the other way around, and he knew I got it from the start. Companies were building structures, machinery, and tools without looking at how the person who was going to work or use the material would interact with the material or equipment. Ergonomic translates from the Greek word meaning "the laws of work," but for Dr. Smith and the rest of us today, it is more like the laws for human activity. Every day, I see situations where some engineer designed a mechanism without thinking about the people who are going to use it.

Just yesterday, I went through two buildings with security systems where they check your identification and give you a card with a bar code, and then you are supposed to use the bar code to get through the gates to get to the elevator. Only, the bar codes don't work, and it's not clear where you are supposed to place them. A security guard must stand there all day, opening the gates. It's silly, and when I pointed it out to the security guards, they just shrugged their shoulders.

I see the same thing with the bike lanes and pedestrian lanes in the city that are so confusing; it makes it more dangerous for pedestrians than it was before they put in the lanes. I see this problem all the time in traffic configurations and get frustrated when I try to use those drive-up windows where people have reached out way beyond their body's capacity to give you your food or take the money. I could cite many examples, but one of my favorites is when I try to get money from the ATM at the bank and end up denting my door and injuring my shoulder just to get a little cash.

Everywhere I look, I see ergonomic problems. The basic idea of ergonomics is for humans to avoid nonneutral postures that will send us to the hospital right now or over time. Lifting more than fifty pounds in an uncomfortable position, like reaching for groceries or reaching for even lighter weights with your arms outstretched, can cause damage. All engineers should take courses in anthropometry to make certain they learn the physical limits of the human body. No wonder shoulder and knee replacements and back injuries continue to be on the rise. In my job as a safetyman, I see situations that will eventually cause them pain or damage. There is plenty of room for we ergonomists to observe humans working and living in a world with so many hazards. Chairs are the worst culprits, but desks with the wrong height will cause pain, and there are also tools like shovels, saws, and hammers that are not far behind. If these tools we use every day put our body in stressful posture, it just takes one false moment, and we can suffer an injury.

THE HUMAN FACTOR: HUMANS MAKE MISTAKES

The human factor is just one of the elements to be evaluated in the field of ergonomics. I have been asked to explain the differences between human factors and ergonomics. I describe ergonomics as the laws of work, but today it is more like the laws of human activity because ergonomics is no longer confined to the work environment. There are numerous aspects to ergonomics, including the capacity of human movements and tolerance, but one of the most important aspects is described as the human factor. The human factor has to do with the predictable behavior of a human being under various circumstances. I have been asked if some injured person made a mistake when they put themselves in danger. This is a tough and important question because maybe they did make a mistake, but maybe the human behavior could have been anticipated in advance, and the hazard and the injury could have been avoided. Everybody makes mistakes! This is called being proactive for safety, and it is apparent that being proactive is not easy for us humans.

The idea of safety engineering is that even if somebody does make a mistake, we might be able to protect them with advanced knowledge and technology. In most cases, we can make it safe by understanding and analyzing human interaction with a tool or a task. It is wrong to blame the injured party for actions that are beyond their control or activity that can be anticipated with human behavior, psychology, and physiology. If we just blame the injured person and don't look at the human factors, the situation might never be improved. Failing to examine the human factor wastes resources, both financial and physical. When we identify a situation as improper, strenuous, or repetitious lifting, we can intervene with material handling, equipment, or techniques and solve a problem not just for this single instance but every time there is a lifting or material handling situation.

They say the definition of insanity is doing the same thing over and over and expecting different results. Well, this is complete insanity. Humans will trip and fall and make mistakes again and again, and if we just pay money and don't change or fix the situation, that is insanity. Historically, nobody has cared enough to examine humans acting in various situations. We think the problem will go away on its own. The amount of time a human driver or machine operator is distracted or fails to follow a safety rule can be monitored and determined, and that information can be used to engineer solutions.

Humans are complicated and fragile creatures, and we are not protected by scales or shells. Our bodies are soft and easily cut or harmed. We make mistakes every day, sometimes every hour. We can't see everything or keep track of everything. It's very difficult for humans to keep track of spatial relationships. We don't have tape measures in our heads. If I ask a group of people how far the power lines are from where we are standing, the magnitude of the inaccuracy of the answers is surprising. Humans are curious, impetuous, easily distracted, and tend to be tired or bored, and this affects alertness, even if they aren't on medications, legal or illegal.

The job of the safety professional is to engineer environments where the limitation of human behavior and ability can be tolerated. This is another book. I have read numerous safety books in my life and career, and none of them focus on protecting people from themselves. I was very fortunate to study under Dr. Kaul U. Smith at the University of Wisconsin in Madison, Wisconsin. Dr. Smith was my major professor and probably the smartest man I have ever known. He wrote a book called *Behavioral Characteristics of Work and Man* that taught me the importance of understanding human behavior. I still use the things I learned from Dr. Smith and his writings every day in my work and my life. He always started by examining the human being in solving behavioral problems. He would examine human history, human capacity, and human capability and then determine the engineered solutions for creating a safe environment. I believe they do this at Disney and Universal and other public family places.

Thousands of cameras and professionals behind those cameras watch the people and intervene to prevent problems. Unfortunately, they do very little of this in the workplace and the rest of the real world.

Any hypothesis suggesting that people are going to look after themselves is a bad start. This hypothesis has been tested, and humans have a poor record of doing the right thing and being proactive for safety. Drinking and driving, speeding, taking shortcuts, using equipment and chemicals without training are just a few examples. When was the last time you saw someone find a tripping hazard and take action to have it corrected? The limited ability of humans to protect themselves and be proactive for safety and health is costing us billions. We need a different approach.

Today's modern world with congestion, aging population, cell phones, computers, terrorism, guns, and danger everywhere requires creative solutions. We must examine and create safe places for people to walk and work, and we can no longer use just design engineers. We need to use safety engineers, ergonomists, and human factors experts. Examining the human experience, they will design places that accommodate human characteristics and limitations, and accidents and injuries will be avoided. When I go to a workplace, especially railroads, steel mills, and construction sites, I find that they don't have safety plans to accommodate their people. The first thing we do is videotape the employees' actual behavior and examine it. Where do they park their cars? How do they walk to the front door? Where do they go once inside, and what kind of habits do they develop when they are working? In one location, we found that workers were sneaking behind a dangerous machine to have a cigarette, so they didn't have to go out in the cold weather. If we find behaviors or conditions that are unsafe, we take appropriate action. In this case, it was putting a heater over the designated smoking area. It's not that difficult if we take the time to care about safety. The first reaction we got from management about putting the heater over the designated smoking areas was that they wanted them to freeze to get them to quit smoking. This reminds me of that commercial where the man says, "It's not that difficult. You eat the food, and you lose the weight." We can do it if we just try. Excuses from management and employees about the protection of their personal privacy are obstacles that must be overcome. I think using privacy issues to avoid looking at human behavior is ridiculous. We can't afford to allow our trained workers to get hurt and lose their service. We must genuinely be concerned about our people and their families and take the necessary steps to be protective of their lives. The safetyman cares.

HUMANS FACTOR AND FASCINATION

Why are we attracted to the accident and injury? We are supposed to be civilized people. The fact is that it is a human characteristic to enjoy a certain amount of violence and danger. We like football, boxing, cage fighting, and NASCAR. I recently saw a video with the horror involved in bullfighting. It is difficult to believe we still allow this torturing and killing of the bull. I am told that people will fill up a bullfighting ring where it is allowed. Like it or not, many people like danger, violence, and even aggression.

We are also fascinated by tragedy. A car goes over a cliff, and we stop, pull over, and look down. We are fascinated by the wreckage. There is evidence that people want to visit sites of great tragedies. They had to tear down Columbine High School in Denver where all those children were killed because too many people wanted to visit the site of so much horror. On the other hand, when someone needs help on the side of the road, and we can really be of service, most of us just keep driving. Human behavior is so interesting.

Many people say they want to help victims, but that doesn't explain the crowds of cars and blocked traffic when a horrible accident occurs. We can't do anything to help, but we still are mesmerized by the situation. We say we are fixated because of concern for the people or a need to help the people, but the truth is that some primitive instinct makes us want to watch. We are voyeurs. Those traffic jams occur not because of the accident (incident) itself but because of the gawkers looking to see what happened. They call it gapers' block. It doesn't make sense to slow yourself down and slow down the rest of the traffic to stare at the catastrophe, but we can't help ourselves.

This happens in factories and on construction sites where someone is injured or killed. The people all stand around and stare. It's not uncommon for management to give everybody the day off because nobody is going to work anyway. The people will tell you they are staring and talking because they care so much about the injured or dead person, but is that what is really going on? These people are stimulated by some primitive impulse to get information and participate in a social ritual.

As a person who has investigated hundreds of accidents, I can tell you that people get in the way. They even make up stories about what happened. They can mislead an entire investigation. Why do they get in the way of the investigation? They are human, and they see everything from their own perspective, and their perspective may be jaded or even unbalanced. Trained investigators know they must interview many people to figure out what really happened and document the facts to prevent a reoccurrence. One of the most important factors in accident investigation is to preserve the evidence because the normal inclination seems to be to remove or destroy the evidence. I have gone to countless sites where, for no good reason, they have moved the equipment or cleaned up the evidence.

MORE HUMAN FACTORS

Before we get to the shopping cart dilemma, more needs to be said about the human factor and human behavior. I like being a human factors expert for several reasons. Nobody seems to look deeply into human behavior, and there are so many mysteries and myths about human behavior. I started seriously watching how people act in college, and I have been doing it ever since with pleasure and terror. There are many surprising things to examine in the culture of an institution and of people.

The corner office is interesting and important, as it represents a nonverbal clue about positions of power. Corners in general seem to be powerful places. When I go on an investigation, I pick out leaders by looking at the corners and the groupings of individuals. I try to assess nonverbal communication by the seats or locations people choose when they are in groups. I have found that I can learn about the culture from photographs, pictures, furniture, neatness, and many other factors. The ergonomics at their workstations tell a story, as well as violations for electrical, housekeeping, and fire protection. These nonverbal communications tell me a great deal about the culture. I could write another book about culture, but for our purposes here, I think it's important to note that people are all different and that generalities are always imperfect.

There are some characteristics of places with cultural problems. These are places where the management wants to dominate, and there are many examples of the company culture involving pleasure, excitement, and risky behavior. I had one client that was a large tool-manufacturing institution that supported car racing and had race car pictures all over the facility. I was not certain they were sending the right message. The extent of the narcissistic culture in an organization is worth examination. If I had to pick a single word to describe people, I would pick insecure.

When people do something unexpected, odd, or antisocial, it is usually insecurity that is behind the cultural problem. I understand that we are all insecure because we are born in mystery in a universe that is not understandable, and we all must face our own limitations and mortality. This is what makes us insecure, and it is bad for solving safety and health problems. I like to watch people and see how they act and how they might act in different situations. Sometimes they act differently than I would expect. I try to think about their insecurities and what makes them tick. It is fascinating to me when people are bossy or mean and even when they are overly friendly. I try to figure out their objectives. There is a woman in a client's office who is nice, efficient, and friendly. She helped me on numerous occasions with my work. I want to thank her for helping me, but whenever I give her a compliment, she acts unhappy. I can tell that she doesn't like the compliment by the way she frowns and walks away. I have not met many people who don't like to be complimented, so I am trying to figure this out. I always say that if there is something I don't understand, it means there is something I don't know.

When thinking about this woman, I wonder if maybe she associates my compliments with some negative action she had with men in the past. She must think that I want something back from her when I give her a compliment. I have learned to just accept her help without any acknowledgment, but I can't help but wonder what is going on in her head. It is hard for me to understand when people don't do things that are in their own best interest, and accepting compliments seems like it would be a good thing. People are flawed and needy and carry so much baggage. When dealing with culture and people problems like safety and health, you need a skilled and experienced person.

I have wondered about strange characteristics that humans have, such as not talking in elevators or always taking the same seat at a table. I like to watch how people cope in different situations. When I was a teacher and had classes in safety every week, I would do things to see how the students would react. I would move their stuff from one seat to another, or I would ask people to separate in different ways, like men on one side and women on the other. One time, I broke them out between glasses or no glasses. Was I picking on people with glasses? People take these changes personally for some reason, and I was just studying their behavior. It is so interesting to see how they react, sometimes with confusion and sometimes anger or just curiosity. This is one of the reasons I like the television show *Survivor* so much.

People are broken into different groups, and you see how they react and how leaders develop and what strategies they choose to try to stay in the game. I especially like tribal council, where they talk about what is going on in the group. I think that at one time, our society operated on the principle we see at the tribal councils. From that show, I learn about different characteristics of people and how they work together or work against one another. I also learn about safety, as people often get hurt on the show. Sometimes they are inexperienced or careless while cutting coconuts or starting fires, and mostly, they get hurt when they work together on tasks, lifting and moving heavy objects and not coordinating and planning to make certain someone doesn't get caught between the parts under a heavy object. A woman contestant almost broke her leg trying to remove a wheel from a wheelbarrow in a hurry; it dropped on her own body. She was in a hurry to win immunity when she got hurt.

I have unending questions about people. I try to understand people who have different political affiliations and beliefs. I almost never get angry because they have a different point of view; I just try to put myself in their shoes and figure out their thinking. Sometimes I can't figure it out, and it is usually because it is something from their past or some prejudice that I don't appreciate. Things that are far out of my range are the most fascinating, like people who intentionally put themselves in danger. I often think that people rarely tell the truth or show their true persona to me and the world. I think people who appear to be happy might just be pretending to be happy. I like to figure people out so I can solve the problem of keeping them safe.

STANDING JOBS (ANOTHER HUMAN FACTOR)

I have never read anything about this problem, but I don't understand why there are so many standing jobs. Standing for many hours in the same place is not ergonomically sound. It's not good for blood flow, and it will punish the back over time. So why do the people at the hotel have to stand all day? Why do the checkout people in the store have to stand all day? I talk to these people and express my sympathy and concern, and I always hear the same thing. Their backs are killing them. When I ask management and workers why they can't have a stool or a chair, I never get a good answer. It seems like they associate sitting with laziness.

This is one of the many things that a safety professional learns, looking at the world from a protective evaluation. They observe many different activities that do not make sense. I suspect that the managers who make their workers stand all day think there is some advantage for the company, but really, in my mind, it is a disadvantage because I suspect there will be more absences, downtime, and even compensation for stress and strain on their bodies. Years of standing in the same place will take its toll.

Also, I have noticed that the desks in front of the standing workers, especially in hotels, are too low to accommodate a sitting customer or too high to accommodate a standing customer, which means it is difficult for the standing workers to even lean at all against the table. I have seen the standing worker needs to bend over to use the computer on the table and lean awkwardly to get the guest receipt out of the printer. It is my feeling that these giant hotel chains have never thought of hiring an ergonomist or human factors expert to examine these jobs. Some evaluation could reap significant economic rewards.

Another standing job that gets my attention is the server inside the drive-up window at the fast-food store. I can't believe how far they lean out that window and reach out to the car to give out the bags of food and accept the money. I have noticed that they do it an average of three times per customer. They give the food, accept the money, and then give the change and receipt.

Every time I see these nonneutral postures, I see nothing but compensation claims. There has to be a better way, and I have noticed the drive-up window production usually causes the standup workers to ask the driver to move up to a parking space, where they are required to run outside, sometimes in the middle of winter, to deliver the food. This seems like a perfect scenario for a slip and fall. I have no doubt that robotics engineers have already developed systems like those used in hospital operating rooms, where levers can be used to let mechanical arms and hands to do the work. There has been angst about these robots and tools, and the public thinks that these systems will eliminate jobs. What I see are these robots and tools eliminating medical costs as well as pain and suffering.

THE SHOPPING CART DILEMMA

Human factors and behaviors remind me of the shopping cart dilemma. My wife and I can't understand why people won't return their shopping carts to the cart corral. It is apparent that most people just don't care about it. There is an engineered solution at a local grocery, where you must pay for a shopping cart, and you get your money back when it is returned. Here is a good example of an engineered solution to a human problem that works. If we could just employ the same kind of thinking to other safety and health problems, maybe we could solve them.

We need people to think about these problems and come up with solutions. In some cities, there is a new mode of transportation called the electric scooter. We can use our credit cards and find one anywhere it exists in the city and ride to the next destination. The only problem is the same as the shopping cart problem, and that is that people refuse to return the electric scooters to their racks. They leave them all over the place and block the sidewalks and the streets, and people run into them and fall and hurt themselves.

Somebody should have applied the shopping cart solution. You must pay for the scooter, and you don't get your money back until it is returned to the rack. The same shopping cart solution could be used in many ways to get humans to do the right thing. It is counter to the disposable society. I can remember when we had to put a deposit on glass bottles at the grocery store, and we would get our money back when the bottles were returned. That was so much better than having people just throwing their plastic bottles out the windows of their cars. This same concept can be used to deal with chemical hazards when somebody buys furniture stripper, paint, oil, or any chemical that should be disposed in a proper manner. They should have to pay a ten-dollar fee that they will get back when they take the material to the disposal site. This will keep them, or at least most of them, from dumping it down their toilet, in a stream, or even in their backyard.

This idea is understanding that humans can be lazy, especially when nobody is looking. The fact that they might take a shortcut and do something that is bad for the rest of us is central to safety and health. It shouldn't be a dirty little secret that they might be unsanitary or careless with their biological fluids. Just go into any public restroom, and the story is clear. Unless that restroom is being cleaned every hour, it will turn into a petri dish. It would not be surprising to find that the kitchens in restaurants are not much different when not reviewed by the health department. I can't believe all these people who are against regulations. Wait until they get sick at their local restaurant. Hospitals are notorious for disease, and let's not even talk about cruise ships. Imagine the frustration on these cruise ships where they are wiping down constantly, and they still get norovirus occasionally. It's a cesspool out there, and guarding the safety and health of humanity is a big job.

FIRST JOBS AND ADVENTURES OF THE SAFETYMAN

MOM WANTED ME TO WORK AND EVEN HELPED ME FIND JOBS at the Shure Save in Skokie and at the Kosher Sausage Company at South Water Market. I learned a great deal about the world and about safety at those jobs. I can remember that some of the beautiful Latin young women at the sausage company would sometimes get their pretty little fingers caught in the machinery, and there was blood, and usually, they would come back the next day and work with a bandage. Sometimes they would lose a part of their finger.

One time, I almost lost my own fingers in a machine. I was asked to cut dry ice with an unguarded bandsaw, I had no idea how to do it, and nobody showed me how to do it. I did not know that the block of dry ice needed to be secured, they call it "jigged," to the platen or cutting table, or it would slide when the blades hit the ice. When those blades hit that ice, the block of ice flew off the table, and my hand was heading toward the operating saw blade. I got a small cut, but I realized that I had just come very close to losing my hand.

The former mayor of Chicago Rahm Emanuel lost a finger while working at a fast-food establishment. I can understand how this could happen to a young person in a dangerous world. I am grateful it didn't happen to me. That sausage company taught me many lessons. A giant African American worker was moving a stainless steel drum filled with animal blood across the aisle, and he asked me if I could hold it on the dolly while he reached for his cigarettes. He handed me this dolly that was pivoted on the back wheels, and it must have weighed more than a thousand pounds. As soon as I started to hold it on the pivot, it came down on the wheels, and I could have broken my back or lost a foot. The worker did it on purpose and thought it was funny.

One time, at the grocery store, I was asked to throw shopping baskets of garbage inside the incinerator outside the store. When I dumped a load, there was a big explosion. The door flew open, and fire flew out of the incinerator, burning my hair and my eyebrows. When I walked into the store and told the manager what had happened, he told me there must have been some aerosol cans in the bins. This was a hard way to learn about flammable liquids and gasses, but it was a lesson that was never forgotten and led to a career in safety.

One summer, I got a job from the *Daily News* to sell Photron cameras. I always remember it was "a camera with one idea in mind, complete simplicity of operation." We were told to memorize a sales pitch, handed a camera, and sent in a bus to some neighborhood. At least twenty of us new hires would get out of the bus with our cameras and our sales pitch, and we would go up and down the block, knocking on doors. Believe it or not, people used to let us in, and once we got in the house, we would give our sales talk about pictures of the kids and the family and that the Photron camera had the best pictures. Once we had them charmed, we would then bring out the contract. I was surprised that this camera cost around $3,000, and I was trying to sell it to people who couldn't afford groceries. This was the first time I realized that people took advantage of other people for gain. I sold one camera to an older lady and felt so guilty I couldn't even sleep. I tore up her contract and went back that night to retrieve the camera. There was no way she could afford it.

The next time they sent us out, we all got arrested for soliciting after-hours and taken to the police department. The district manager had to come and get us. That was the end. I had another job selling Fuller brushes and brooms. I did well at that job, which also involved going door-to-door. I learned that the idea of cleanliness and good housekeeping was helpful when selling these products. My best seller was a stainless steel sponge for cleaning pots and pans.

The only other job besides OSHA where I learned a great deal about life and people was when I was a cab driver in Chicago. I learned that I had to bribe the dispatcher to get a decent car, bribe the helper to make sure there was gas in the cab, and, most importantly, bribe the O'Hare ticket taker for short trip tickets, so I didn't have to spend the rest of my life waiting for business. This must have been around 1972. I felt sorry for the other cab drivers who were from different cultures and didn't know how to work the system; they had a hard time making a living. I made almost $300 a day from tips and kickbacks at restaurants, strip clubs, and hotels. Those jobs were important to the development of my life and career and helped me to understand things that led to a career in safety and health.

I have an affinity for working people. It may have come from my father, who was a working man, or it may have come from my working experiences. I saw how hard life could be for people. Some of my associates like to hang out with one another in their private clubs and organizations. I have never been big on private clubs and organizations. I like getting out in the world for stimulation and the excitement of learning new things.

To me, the working world is the real world. It's difficult for me to see people of any age who don't enjoy working or who think about retirement and their own pleasure. I don't want to sit on a beach or at a country club. If I didn't have this wonderful safetyman job, I would still love to drive a cab. I would enjoy working as a greeter at Walmart. I would like to try some jobs I have never done, like being a server in a restaurant. I just like working, and I like people. When I see signs in the window for help wanted, I think, *I would like to take the job.* I would work at a fast-food restaurant in a minute for the experience. It would be fun to learn the fast-food job, and I would be especially interested in serving people in the drive-through lane.

One time, eight years ago, the city of Chicago was looking for a safety engineer to do some work evaluating schools and fire departments for safety and compliance with the LEEDS environmental program. I called them up and worked for them the whole summer while I was still doing my consulting business. I had to go to city hall every day, so it was a long trek, but it was different and rewarding, and I learned more about Chicago politics than I did about compliance with the LEEDS program. I have suggested to many people that they should have a job or even two jobs to get them out of bed or out of their chair and into life. People think of jobs as just about money, but I know that working is about life, not about money. Sometimes the people I suggest getting a job just sit at home watching television, and eventually, they get into some trouble or lose their health. Work is the secret of life, my friends, and good work is a great blessing.

ACCIDENTS AND INJURIES

How many accidents and injuries can you afford to have in a lifetime? I try to get people to think about their experience in this world and get them to spend more time preparing for danger. Many people tell me they think planning for danger is a waste of time. They either think their fortune is out of their control or that planning will not make a difference. I try to get them to understand. I see all these accidents, injuries, and deaths and how they could have been avoided. There are exceptions like an act of God. There could be an earthquake or a tornado, but aside from natural disasters, there is always something that could have been done to avoid the incident.

I have cases with equipment failure, explosions, overpressurization, fires, radiation exposure, chemical exposure, and all could have been avoided if the hazard was eliminated or people were kept away from the hazard. Testing the operations of processes and equipment before they are approached by workers would save countless lives. There is a testing method used in chemical plants and refineries called "what-if scenarios." Using these methods for each process and procedure, the safety professionals must consider every possibility of what could go wrong and develop a method of safeguarding under those circumstances. OSHA's Process Safety Management Standard requires the development of redundant controls and safeguards for the protection of workers under any possible conditions.

There was a television program recently on the Chernobyl disaster that occurred on April 26, 1986. This was the worst nuclear incident in history. The incident was the result of flawed reactor design and poorly trained personnel. There were as many as thirty thousand people near the reactor when it exploded. The greatest harm was to the firefighters who were sent into the exposure, and there was no containment building. This is an example of not enough contingency planning for the possibility that something could go wrong. Had firefighters had more protection and radiation suits, the losses could have been reduced.

The point is that planning for something that can go wrong is extremely important work. We might not solve every problem or eliminate every disaster, but we can still have a big impact. If people spent more time thinking and planning for the prevention of accidents and injuries, we would have a safer planet and society. The internet and technology are helpful in these matters, but we must have the resolve to make planning for a disaster a higher priority.

ACCIDENT-PRONE CONCEPT

The idea of blaming the injured person is an unfortunate aspect of safety. If we just blame the injured person, we can avoid all the hard work of solving the problem and getting to the root cause. We already know that humans are fragile and flawed. This issue has raised its head in the concept of being accident-prone, which has been studied and discussed for decades. Someone accident-prone is involved in a greater than average number of accidents. There have been studies that have found that 27 percent of people have more accidents than other people. Youth, inexperience, and dissatisfaction are characteristics of injured people. They also have found that physical problems, such as a sleep disorder or smoking or not being physically fit, are issues. They even found that people without a hobby were afflicted. One study found that people who were open, dependable, and agreeable were safer than average people.

I don't believe in being accident-prone. From my experience, some people may have repeated accidents and injuries, but the main idea of being proactive for safety and health is that the workplace (where these studies are conducted) and the world must be designed with human fragility and limitation in mind. This is the basis for an understanding of ergonomics and human factors. I believe that people and organizations that are thinking about people being accident-prone are living in the past.

When people go to a hospital, amusement park, or concert or drive on the road, there is so much opportunity for disaster. There are thousands of situations where people of all types are invited, and there is little planning for potential injuries and illnesses. These places must consider people who are sleepy, not physically fit, in wheelchairs, and utilizing crutches. They must even accommodate small children. There are certain elements of luck, and it is true that some people have more unfortunate circumstances than others. The world is complex, and the interactions between people are always going to open the possibility of random actions or even acts of God. I have been asked if I believe all accidents and injuries can be avoided, and I believe they can never all be avoided. There are too many circumstances out of our control, such as tornadoes, earthquakes, cyclones, fires, and many other hazards. On the other hand, I believe 90 percent of the injuries and illnesses can be avoided with planning and protective action.

The important and interesting thing about being a safetyman is this interaction between accidents (incidents) that are caused by things outside human control, accidents caused by the actions of individual people, and accidents caused by a lack of planning and compliance with standards, customs, and practices. Looking into these causes, solutions, and relationships is a great challenge and opportunity for satisfaction in our work.

CONFIDENCE OF A SAFETYMAN

Once you have been the safetyman for many years, people know it. You get a level of confidence. You are doing a job that's hard to do, and others won't do it. After consulting for many years, people had heard about me and watched me, especially during accident investigations. I try to see the big picture. I was involved a case where a truck driver was crushed by a bundle of pipes that rolled off his truck, knocked him off the truck, and crushed him on the ground while they were being removed by a crane at the delivery site. I noticed the rolls were rounded and saw that the bundle of pipes had rolled when another bundle was being removed. Other experts testified the bundles could not roll, but there was no other explanation for why they hit the driver and knocked him off the truck. The bundle landed on his legs, and he lost his legs. I argued that the rolling motion was the only possible answer. The confidence I had in my knowledge and abilities allowed me to testify and convince the jury to help the family of this poor man. The jury concluded that both the company that formed and loaded the pipe and the company that lifted the pipe off the truck were at fault. They could see that the bundle had rolled and was not properly stabilized when loaded and unloaded.

Another time, a concrete cornice fell from a roof upon a worker. I suspected there wasn't enough rebar or support for the load. When they were working with vibratory equipment for the demolition, it caused those capstones to come down and crush the worker, who should never have been allowed under the load in the first place. In too many cases, it is the lack of support for structures that are supposed to be secure and stable that causes a severe injury or fatality. One time, at a church, a statue of Saint Joseph fell over and crushed a church volunteer who was painting the structure. This person was holding on to the statue while painting the face when his weight caused the six-hundred-pound statue to fall on him.

In these cases, and most cases, it is crucial for the safety professional to have the confidence to see and report the science and the truth. As safetymen and safety women, we have a safety hierarchy that is important and not well understood by the rest of the world. The safety hierarchy idea is that the most important concept in safety is to engineer the danger out of a capstone or a statue. It is only when the danger cannot be engineered out of the picture that we turn to guarding or warnings. Many laypeople think warnings are all you need. In the end, many of these cases are decided by laypeople who serve on a jury. They are the regular people who are not biased and not engineers or specialists. My experience is that if you tell them the truth about the safety hierarchy, they will recognize it and do the right thing.

Confidence is not something you get overnight. To gain confidence, there must be experience and hard work. There is no substitute for confidence when you are trying to do the work of a safety professional. Today I have the most confidence. My office is like a safety cave. I answer my telephone, "Safetyman, here I am!" I have quality clients, and I can turn down any job if I don't feel comfortable. I have reached an important milestone of being in a position of total independence.

When I am called to an accident or a fatality, I listen carefully and ask a lot of questions. I want to know if there is a legitimate violation of standards, customs, and practices. I look at all aspects of the incident and remain objective and open to the facts. It could be the injury or fatality was as an act of God or nature, or it could be the injured party's own fault. If the use of alcohol or illegal drugs is involved, it can change everything. I look at both sides of every case when deciding whether or not I want to be involved. I also look at the quality of the people who hire me. I will work on a case that supports the standards, customs, and practices. I have seen unfair cases where I thought an accused party was innocent and shouldn't have to pay damages. I don't need to pick any side. I just need to represent safety and health. I have run into too many safetymen and safety women who are insecure representing safety and health. As soon as they start compromising to benefit one side or another, they are ruined and cannot be a true safety professional. This is the hardest part. I tell people that I represent safety, not one side or another. Safety needs to be represented, and who better to represent it than the safetyman.

THE MORAL COMPASS

Safetymen and safety women have a special responsibility to look out for other people. I have stated that this is a difficult job and an awesome responsibility. Often, the safety professional must go against the grain and stand up against powerful forces that want profits and production and people who don't believe in or care about the safety and health of other people. A safety professional without a good, strong moral compass can get into trouble.

I was brought up with the belief that anything that was accomplished would be accomplished with hard work. There were no shortcuts. My father and mother believed there was nothing more important than hard work and that the harder they worked, the greater the rewards. This upbringing allowed me to feel fearless in the face of opposition. I always had the feeling that I could work hard standing up for what I thought were the right and moral things, and I could not only survive, but I could make a difference. I now call this thinking the moral compass, and I didn't realize how important that compass was until working many years in the safety and health field.

I noticed that some of the safety and health professionals were not sticking to their principles, or sometimes not to the standards if it meant they would have to have conflicts with management. I could see as time went by that not selling out is so hard. There are so many more opportunities in management and production than there are in fighting for planning and preparation. Almost everywhere I went, people were telling me that safety was slowing things down, and all these rules and regulations were just a burden. I had the impression and continue to believe that to make it to management positions, I would need to learn to compromise more and work with the management team. What I found myself doing was asking a lot of questions and making people prepare and plan for each activity, and this was not making me a popular person. I can't emphasize enough how important this is in being a safety professional.

Every one of us will have this moment of truth occur again and again where we need to decide if we are going to continue to be a pain in the butt to people who don't care or understand. It is difficult to get management to worry about things that have not happened yet or that may never happen at all. Management will often tell me that they have been doing something one way for forty or fifty years, and nothing has happened. I tell them something bad will happen when the equipment is not locked out or when the guard is not in place, but they often don't believe me. This is the everyday life of a safety professional, who seldom makes it into the top management. The safetymen or safety women sacrifice opportunity and upward mobility.

The situation has improved over the years, but to this day, the top jobs in management are not safety and health related. Money is an important objective, and without money, life can be tough in our American society, but compromising on requirements for safety and health is too high of a cost to bear. I have always been independent enough not to feel too much pressure to abandon my post. I was never a good ass-kisser, and sometimes it got me in some trouble. Sometimes I felt that if I had done a little ass-kissing earlier in life, I would have had a more comfortable life, but I couldn't be the safetyman. Looking back, I have no regrets. I remained independent, and I have had a good life. I feel like I have done some good work going against the grain. My independence has brought me excitement and success. It's a difficult task, but a safetyman *has to* stay independent. When a safety professional loses their independence, they cannot do their job. My clients have been surprised that I don't always try to please them. I represent safety and health. I tell them the truth, including the things they might not care to hear. I don't knuckle under the pressure for profits, production, promotions, or public relations. I call those the four Ps. I managed to work at OSHA for almost eighteen years and for hundreds of clients and corporations, without sacrificing my integrity or giving in to the pressures that exist everywhere.

SAFETY INCENTIVES

INCENTIVES SHOULD BE USED TO REWARD PEOPLE WHO WORK hard on the safety and health program. They should never be rewarded for not having injuries or illnesses. I am not sure where the idea came for rewarding workers who don't get hurt or sick, but it tends to just make them feel bad and hide actual injuries. I have seen people who have been hurt on the job working in casts or even on crutches; rather than reporting the injury, they try to hide it to get some bonus or reward. I have also seen companies who create a light-duty job for workers injured on the job to avoid reporting the injury or illness to Worker's Compensation or OSHA. We must look at injuries and illnesses right in the face and find out what caused them and how they can be avoided. Any incentive program that rewards for not having them is a problem for the entire effort.

LOTTERIES, HORSE-RACING GAMES, AND OTHER TRICK OF THE TRADE

Sometimes these safety incentive programs take the form of games or lotteries. Signs are posted, bragging about how long it's been since someone had a lost workday injury. These are tricks to put safety responsibility on the employees or the workers, exactly where it doesn't belong. These programs don't work and often work against the idea of accident prevention. These incentives tell the workers to not report their injuries, or else they will lose their reward, their lottery ticket, or even worse, they will get the whole crew angry because they all lost their pizza dinner.

For the safety professional, the idea is to examine the workplace and worker behavior to prevent accidents and injuries in the first place. There is nothing wrong with taking the money saved by a good safety effort and sharing the rewards at the end of the year with the entire staff, but doing this weekly or monthly can have the opposite effect. I remember working for a large tool company with plants around the country and seeing workers with bandages and even slings working at their machines. I asked to talk to those workers and found out that they thought no matter how bad their injury, they could always work "light duty," get paid, and contribute to the company incentive program. The whole idea of light duty can be confusing because it is a good idea to get people back to work as soon as possible, but having them work with casts and crutches is just a way to hide injuries. The way injury statistics are collected, it is essential that employers and businesses report injuries consistently and accurately; otherwise, we have no idea where to look for problems. When companies come up with ways to hide injuries, it sends the entire system down the drain. I believe every lost-time injury should be reported immediately to some computer base. If a person receives an injury that prevents them from working, it is a very serious event. If workers are encouraged to work from home in any condition at all, it defies the entire system. This working from home should be examined more closely.

RESPONSIBILITY FOR SAFETY AND HEALTH IS THE MANAGEMENT'S

People say that safety and health is everybody's responsibility, and it is difficult to argue with that premise. The real question for the safety professional is, who is in the best position to assure the principles of safety and health are implemented? Who can hire safety people, buy safety products, and implement planning, coordination, and procedures for safety? Management must do it. Each person should participate in the process to follow rules and report hazards, but they are not able to examine the big picture and make substantial changes in the culture. One of the worst examples of putting the responsibility on workers for safety and health are the incentive programs mentioned above. These programs act like accidents or injuries are the responsibility of the workers. The programs encourage workers to hide injuries and sometimes involve gambling and lotteries. One of the biggest incentive programs provides prizes and lottery tickets for no accidents or certain types of behavior. It's not the workers who are responsible for the safety programs, and each employee has a basic right to believe that a safe world will be provided by management or government.

RESPONSIBILITY FOR CONTRACTORS

Some managers think their safety program efforts are only for their own employees, and they have no responsibility for contract workers, temporary workers, or anyone else that enters their facility. They are completely wrong, and sometimes they don't want to hear it. The top management of businesses and construction operations are the only party in a position to make certain the safety and health program is implemented in a way to protect everyone exposed to hazards from the work being conducted. They are the only ones who can make certain that one company or contractor does not create hazards for another, or that one contractor or temporary worker doesn't make a mistake that causes a catastrophe. Management must take control and assume responsibility for the safety and health effort for contractors, temporary workers, visitors, and others, but it has been slow in coming. I have had discussions with other safety professionals who tell me when they hire day laborers, those day laborers are represented to be fully trained and capable. They might even have contracts that say the same thing. The problem with this is that these day laborers or contractors might be well trained, but they have never visited the controlling contractor's facility, and they have not worked with or near their tanks, chemicals, and machinery. There is no way those day laborers or contractors can be prepared to know how to work in a safe manner at the controlling employer's facility or project. They might open the wrong cabinet or turn the wrong knob. Many companies have liaisons who work with contractors and temporary workers until they get acclimated. This is a good step forward, but those same companies need to acknowledge that they are just as responsible for the safety of those contractors or temporary workers as they are for their own workers.

Today, when I hear a general contractor, construction manager, or company manager say they believe that the subcontractors are responsible for their own safety, I immediately think that they are behind the times and that they are probably paying the price for unnecessary injuries and illnesses. The single biggest change and improvement in the safety and health field will come when these powerful parties who are paying for the work stand up and take responsibility for the safety and health programs, policies, and procedures at their facilities. This will make an enormous difference.

SPEAKING OF THE LOTTERY: THE CANCER LOTTERY

No, I am not talking about the tickets you buy at the corner store. I am talking about a much more important lottery than the one where you can win hundreds of millions of dollars. I am talking about the asbestos lottery, silica lottery, blood-borne lottery, tuberculosis lottery, influenza lottery, and many more dangers in our complex world. We have been exposed to all this stuff, and for some reason, it makes other people really sick, and sometimes they die, but so far, we are still living. One of my favorites is when I talk about asbestos or silica in my classes. I tell the students that each of us has asbestos fibers in our lungs and scars on our lungs from silica exposure. Some people have a single asbestos fiber that hooks into their lungs, and their body's immune system tries to remove the fiber; a scar forms, and over many years, that scar can get bigger and bigger. As with the case of mesothelioma, that scar tissue can take over the lining of the lungs, and it will significantly shorten your life.

With silica, it's similar. Many compounds on construction sites cause clouds of dust that contain crystalline silica, which can cause silicosis. Found in soil, sand, granite, concrete, rock, and many other materials, it can result in severe lung disease and cancer. When I see those clouds of dust in construction and even on the highway when I am driving my car, I think about how each is a dangerous silica exposure that will shorten a life. I often wonder if the average person realizes how dangerous a single exposure might be. Silica is like glass, and when you inhale glass, it will scar your lungs; inhale enough of it, and you will have difficulty breathing. Of course, heredity, smoking, and general health are all factors, but when you think about it like this, you start to realize that we are all playing the life-and-death lottery all the time, and it is scary.

Asbestos is not one but six naturally occurring silicate minerals, which all have long, thin, fibrous crystals. Some of the fibers are visible, but there can be millions of microscopic fibrils that are invisible to the human eye. Breathing any air containing asbestos fibers can lead to cancers of the lungs and chest lining, and asbestos currently kills around four thousand people a year in Great Britain, where they keep records. Believe it or not, asbestos was once marketed as a "magic mineral" with nothing but positive benefits. Asbestos can still be found in many homes and schools, in shingles on roofs, insulation, flooring, sprayed-on ceiling materials, and textured paint, as well as around boilers, piping, and electrical equipment and in appliances, fireplace logs, and automobiles.

SAFETYMAN AND CAMERAS

I was so closely associated with cameras at the OSHA Training Institute that they called it Burg Vison. Yes, I love the cameras because they can watch what people are doing. As a safetyman, it is my job and responsibility to monitor performance and behavior. Some people accused me of being against privacy. I am not against privacy in the bathroom or the bedroom, but in public places, I think we should have video cameras everywhere. According to recent reports, Beijing has closed-circuit cameras (CCTV) that cover every part of the city. London is second, and Chicago is third. In the 1990s, I was looking for a camera in the workplace to identify unsafe behavior and safety and health problems. Everyone thought I was Big Brother. Now these cameras are everywhere, and I think it is a good thing.

I can remember talking in my classes about how a camera creates a safer environment, and my students would come up after class and complain they would get caught sleeping or smoking or flirting or even drinking on the job. I didn't know what to say because that was really the whole idea. I would love to have a camera that would pick up every driver who throws cigarettes, beer cans, or condoms on my front lawn and be able to put them on the internet. My original idea was to put cameras in the factory and on construction sites so we could watch and identify problem behavior and act before anybody got hurt or killed. The idea was that the camera costs around one hundred dollars, and the injury could cost millions of dollars. It always seemed like simple economics to me.

Still to this day, I find myself looking for cameras where they should be and not finding them. I belong to a health club, and I tell management that they need cameras to protect the club. So far, the only cameras they have in place are the ones to take a picture for identification on the membership cards. I wish they would just take the same cameras and put them high in the lobby or on the workout floor so they can see what is going on. I think we know these cameras are important for improving our safety and security. Beware of the people who are saying things like "Big Brother" and that the government is taking over our lives, because I think those people might have something to hide.

SOMETHING TO HIDE

It's not just in classes at the OSHA National Training Institute where people worried that someone would catch them drinking, taking drugs, or doing other dangerous behavior. These and other types of aberrant behavior exist throughout our society. There is road rage. We see it when somebody cuts us off, and I must get the finger at least once a week because I drive the speed limit. It is undeniable that people today seem angrier about their lives. More traffic, more problems, and more people are in a hurry. Too many people are not honest, and they are not living by the golden rule. We need cameras to slow people down, and if they kidnap someone or commit a crime, we will have a chance to catch them because of cameras.

I believe that one day, everyone will have a camera on their bodies, not just in their pockets. These cameras will record lives and adventures but also any threats or dangers. I am sure the footage will be used to find criminals, but I am not sure the footage will be used to solve problems that caused the need for the cameras in the first place. We shouldn't have anything to hide from these cameras. My father was a straight arrow, and the apple doesn't fall far from the tree. We are a family of straight arrows, and that makes us good safety professionals and great lawyers, as in the case of my two younger brothers.

SPEAKING OF CAMERAS AND SATELLITES

We have Goggle Earth. Shouldn't they see this horrible destruction of our planet and act to protect our beautiful earth? They don't seem to be doing enough. This excellent and new technology should shine the light on the bad things that are going on all over the world. I haven't seen the deforestation in South America or the blithe in Africa, but I have seen the diminished glaciers and the deforestation of remote areas of Alaska. It is at least now possible for us to see what is going to happen to our earth, and maybe one day, we will be able to do something about it.

Camera and satellites have been helpful with safety programs. Finally, companies like FedEx and UPS put cameras behind their vehicles, and many children have been saved. Most cars and trucks have cameras when we buy them. They need to put cameras on backhoes, dump trucks, and other construction equipment. I am told that Europe and some other countries won't accept any vehicles or construction equipment without cameras. It is outrageous these companies don't provide a hundred-dollar camera that could save lives and prevent injuries and heartache. Every time I get a case where somebody is run over, it could have been avoided with the installation of an inexpensive camera. A moving piece of equipment can't see in the blind spots, and there is simply no reason not to provide a camera. I don't even like the signalman or flagman idea that is so common in the construction industry. There are too many cases where this signalman wasn't paying attention or made a mistake, and someone got hurt or killed.

Cameras are like eyes wherever we need them. We can see and make certain it is clear before we move. There is no place where cameras couldn't improve safety on the railroad. Unfortunately, they only use them in the parking lot to protect the property and not for railcar or locomotive movements. This is so simple and so cost-effective. I heard a huge profitable company bought a railroad a few years back, and I thought to myself that it wouldn't be long before a smart business person would put cameras all over their railroad to save hundreds of millions of dollars. So far, I am still waiting.

UNSPEAKABLE

I was reading about the Radium Girls. These were female factory workers who painted watch dials back in the late 1920s. They were told it was completely harmless, low-level radiation. It was a lie. Radium has a half-life of 1,600 years. Many of these women became sick and died at an early age. I was thinking today (2020) that things might even be worse than they were decades ago. Here in Chicago, we have companies exposing their workers and the community to ethylene oxide. The facility uses ethylene oxide to sterilize medical equipment and food products. Studies showed an unusually high cancer risk in the area around the company's emission of the gas, and recently, a study showed that the steel plants and chemical plants in Whiting, Indiana, are spewing lead and other toxic materials, not only in Indiana, but it's also blowing toward metropolitan Chicago. They found at least thirty swans with elevated lead levels, and further testing has been ordered by EPA. Whiting probably has high lead levels. Lead is associated with birth defects, brain and organ damage, and learning disabilities.

I have seen people in trucks dumping barrels of what I believed were toxic materials into lakes, rivers, and streams. Even some neighbors throw toxic things in their regular garbage. Two years ago, a neighbor decided to burn mattresses in an open fire in their backyard, and our eyes were burning for days. Lucky for us, the county came and fined them. This is the shopping cart deal all over again. It would make sense to return the shopping cart to the collection area, but people don't do it unless it costs them money. The people who are manufacturing, handling, and using toxic chemicals should have to pay in advance and only get a refund when they prove they disposed of the stuff properly.

Update on ethylene oxide. They have now found it in another county. They are looking for it in other places after more people appeared with cancer in other suburban areas. There is little doubt they will find cancer-producing chemicals in many other places. They have been putting chemicals into our air and water since the beginning of the industrial revolution. We now have the knowledge and technology to find out what damage has been done. There are hundreds of thousands of toxic, cancer-producing chemicals and millions of people who have developed cancer. I would not be surprised if someday it is discovered that many cancers and much human suffering were caused by these chemicals. Yesterday, I was behind a large diesel truck spewing black smoke. I could taste that smoke, and it made me choke, and for a few minutes, I found it hard to breathe. I know that black smoke is a carcinogen. All these companies will say that they didn't know they were causing so much cancer and human suffering.

TALKING ABOUT SAFETY

When I talk to people about safety, either individually or in front of groups, they think it is a boring subject. Sometimes, especially if it gets technical, I can see the lack of interest. But when I talk to them about how it will affect their families, they become more interested. This is called the reptilian brain. This came from a brain theory in the 1960s where the structure of the brain includes a basal ganglion where basic function, such as aggression, dominance, territoriality, and ritual, exists. The idea from a safetyman perspective is that if you can reach a person's primitive instincts, like protection of their lives and the lives of their families, it will get their attention. My experience tells me this is true. When I talk about technical regulations or standards, there is little interest. I believe this is the reason I found it so difficult to teach safety in the thousands of classes I taught at OSHA and in many locations across America. If I didn't give those students a reason to pay attention, they would fall asleep.

In the training business, we rely on audience participation and demonstrations, but I found myself relying on the reptilian brain to gain their interest. Audience participation is great when they are doing something enjoyable, like an experiment or a presentation. I would require each student to make a presentation on a topic, and it kept them awake. The trouble is that that was just one portion of the course, and the rest was a lecture, and they were bored. The next choice is the demonstration. My favorite demonstration was when I would blow up a five-gallon water bottle or a glass tube. The whole class would pay attention, and at the end of the class, when I got my evaluations, I found out that they loved that part of the class the most. I also found out that they liked me better when I blew something up. I used to tell fellow instructors at the OSHA Training Institute that I would like to blow something up every hour, and it would be the best class. I took every opportunity to blow up those five-gallon water bottles, and my evaluations for training got better and better.

The only other tool I had in teaching safety and health was the reptilian brain. I found that this worked too. If I could get those students to think about their primitive instincts and protections for themselves and their families, they would pay attention. I do the same thing today when I speak in front of groups, lawyers, or juries. I try to tell them why it is important for their own lives. The closer I get to the survival mode in the students, the more interesting the material. One thing about safety and health training is that it is a lot of math, physics, and chemistry, but at least there is a protective reason for the course. One of the problems with the courses is there is too much about getting an OSHA card or getting some authorization, and there can be too much emphasis on getting through the topics and meeting the specific criterion. This can be boring for the students, but as soon as you talk about their mortality or the protection of their families, they tune into the conversation.

It is one thing to tell them that a mixture of chemicals, such as drain cleaner and bleach, can be toxic, and it is another to tell them that they may have those same chemicals under their sink, and when they use them to clean their toilet, the resulting chemical reaction, vapors, and gases might end their life or, even worse, the life of their wife and children. This aspect of talking about safety and health is what makes it so interesting, and I have seen some incredible instructors mesmerize the classroom with audience participation, demonstrations, and speaking to the reptilian brain. The more of this, the better.

Now the question about retention of information is a different matter. Each adult student brings a different amount of knowledge with them when they take the training. Some students are far more attentive than others, and the only method available to determine retention is through testing. Nobody likes to be tested, and that part of the course gets an unhappy response from most students. The biggest problem is knowing if they will remember what they have learned when they really need the information. Some safety person might hear someone gaging for air in a confined space. Did that student learn enough in my class to know never to enter that space without supplied air? If my student enters the tank, they will suffocate without air. Did they learn enough to not panic and die? Not enough of my students and associates think about the training and teaching in terms of emergency preparation. Emergency drills and preparedness are more important in situations where there is a shooter in a school or business. Instead of just having fire drills, schools and businesses are now having preparedness drills for possible active shooters and terrorism. More of this needs to be incorporated into safety and health training. If there is a toxic chemical release, emergency, or evacuation, including violence in the workplace, the safety and health professional will need to know the right actions to take to identify the problem and secure and protect the area.

ADVENTURES OF
THE SAFETYMAN

BEING A SAFETYMAN HAS BEEN A GREAT ADVENTURE. I WAS just asked to help in a case where a twenty-two-month-old infant fell into a retention pond and drowned. This infant drowned where the water from the property was draining into the pond. He was with his two-year-old brother. Nothing but sorrow for the family, and the defense attorney is going to say that the mother wasn't watching her children closely enough to keep them safe. She had two other children to watch, and this pond was close to the house. There was a hole caused by the erosion of the water draining from the bank of the pond where the infant fell into the water. The human factor points to the fact that it is difficult for a busy mother to keep track of two small children. She thought her husband was watching them. It also points to the fact that someone should have known the pond was dangerous and should have protected this area of erosion. The mom has only one set of eyes, and the husband said he never heard her request to watch the children. This is going to be a tough case, but I want these ponds and hazardous areas to be evaluated by a safety professional. There is nothing I can do for this family. I can only try to protect another family in the future.

Was it the property owner's responsibility to protect his pond? Is it the custom and practice to fence the pond? I don't think it will matter much in this case, because they placed the fence for protection in the first place. We may have all seen ponds without fences, but this one right next to the homes needed to be protected. This case is worth a look because it hurts our hearts when a young child drowns, and it could be easily prevented. I need to determine the custom and practice for guarding retention ponds. I fear that they are not protecting these ponds, and it could be time for such ponds to have the same protection as swimming pools. No family should have to face the devastation of losing an infant. I haven't investigated the codes and standards yet, but my gut tells me that near a residential housing project, it should be anticipated that infants and small children might wander near the pond. I know there are strict standards for fencing for outdoor swimming pools.

I have had some experience working on ASTM (American Society of Testing and Materials). I serve on both the E34 safety standards committee and the F10 recreation and sports committee. I have helped write rules for swimming pool safety but not for retention ponds. The F10 has also allowed me to work on standards for sports. I had the pleasure of working with the NHL (National Hockey League). This case, like many of the other cases that I see, makes me think about the need for more standards in different areas, which is a never-ending process. I know there are many rules about fences for swimming pools because so many kids have drowned. I wonder how many kids have drowned in retention ponds. I read just a few days ago that a celebrity lost his child in a swimming pool, and I remember that a colleague lost one of his children in a swimming pool, and that was one of the motivations he had for being a safetyman. Every case is different, and each case needs to be investigated on its own merits. Do we want property builders and owners to protect their retention ponds with fences so small children cannot get in there and drown? I am a safetyman, and I vote yes.

Each day, I get calls about catastrophic injuries and deaths. Too many cases, but there is a shortage of safety professionals out there, and it is a matter of supply and demand. I wish there were more safetyman and safety woman. It is not possible to work on every case. There are two sides to almost every story, and I am always looking for good defense cases. The problem for me as a safetyman is that sometimes I don't see any defense. How can I defend a client where there is no safety effort and they are violating standards? I tell them to settle the case before they spend a lot of money on lawyers and experts.

Insurance companies send me cases as a consulting expert, and I give them advice on whether to litigate or settle the case. They may or may not listen to my advice, and sometimes they spend millions and lose. Some insurance companies settle their claims without litigation, and this saves the insurance company the costs of lawyers and experts that are more than the cost of the claims. This sounds right to me. The average litigation is at least half a million dollars, and if they can resolve it for less, it makes sense to me, even if it does mean less work for the safety professionals. The question for me is, did they remedy the hazard? It's good when the cases are settled but only when the hazards get corrected.

I have had a few cases where it seemed like the plaintiff would win and the plaintiff got zero. In one case, the jury disliked the plaintiff's lawyer, and in another, the injured person was unsympathetic. There was one that was quite odd. It was a case where an artist went to a private party at a residence that was an artist's retreat. This retreat was at a house that had a huge sculpture in the backyard. This sculpture was the size of a house, and it was illuminated in the evening. Surrounding the sculpture were lighted platforms around six feet high. They were drinking wine and eating cheese, and the plaintiff met a woman, and there was some dancing. He was showing off for the woman, and he climbed the platform and was standing on the illuminated glass platform when the glass broke, and he fell into the platform. You would think that he was going to fall six feet to the bottom of the platform, but he fell thirty feet to the ground because there was a secret storage area under the sculpture. After falling thirty feet, he landed on a concrete floor within a cave. His injuries were significant. I went into this secret, large underground storage area. No one would ever know it was under that sculpture or platform. To me, this was a hidden defect case, and I was expecting a good resolution for a man who would need medical treatment for the rest of his life. This defect was so hidden, and there are specific standards for skylights, requiring them to hold the weight of any anticipated load. To me, it would be anticipated that people, especially young people, would climb up on these platforms, and that is how I testified. It is never possible to know how these trials will turn out, and usually, the verdict is not revealed until the next week. I found out the plaintiff got zero. I thought to myself, *This isn't right and should be appealed.* The lawyer told me there would be no appeal. That case and that trip was quite an adventure. On the way to the sculpture, I went through the town of Roswell, New Mexico. That place has the UFO museum, and I stopped at that museum on the way to the sculpture. Little did I know that the case was going to turn out to be one of the strangest adventures in my career. Maybe there is something to all that Area 51 stuff.

Speaking of unusual adventures, there was the case at the University of Chicago, where they invented the atomic bomb. There was surprisingly little security or radiation detection. It was a long time ago, but that radiation lasts a long time. An elevator fell to the bottom floor and injured a worker. Elevator cases are all too frequent. People don't realize how dangerous elevators and escalators can be. There were two escalator cases where young children got their feet caught into the side of the escalator and had severe injuries that will last for the rest of their lives. People are always trying to catch an elevator, and they stick parts of their bodies in the door. One time, I saw a father ridicule his son for not sticking his foot in the door to stop the elevator from leaving the floor. These people need to understand more about microswitches or sensors in the doors of these elevators. Sometimes they don't work, or they break down or wear out, and when you put your arm or leg (or that of your son or daughter) within the door, the elevator might leave and take the arm or leg with it. This has happened many times, but a safetyman doesn't know how to get people to stop exposing themselves to this danger.

Yesterday on the train, people kept getting their bodies in the way of the door to the train and slowed down my trip home. These trains have taken off with somebody's limb. Some years ago, a woman with a Stradivarius violin lost her leg on one of the commuter trains but saved her violin, and it was a big story. There are many dangers associated with public transportation, including being hit by a train, slips and falls, and door dangers to the public. There have been dozens of cases where the car cleaners get hurt, but nobody seems to be doing much about it.

Recently in Chicago, someone was waiting for the bus, and the shelter station at the bus loading area was rusted and came down on the commuter, and she subsequently died. Isn't anybody inspecting these shelters? Nobody cares until somebody gets killed, and then they might only care for a little while. The management isn't being proactive and thinking about what could happen when there is not enough maintenance. Too many injuries have been recorded, but too often, they take no action to deal with the root cause. Sometimes it seems it is only the people who get injured who finally understand and get converted to be a safety person. The problem for the safety professional is how can we teach people to be more proactive about hazards and more protective of their exposure to danger?

I have had too many cases where a crane operator or other heavy-equipment operator made a mistake that injured or killed another worker. I've had at least fifty cases where an operator ran over another worker on the jobsite. Almost all these injuries and deaths involved a lack of planning for safety to make sure one worker didn't create a hazard for another. There is not enough research into this area. There should be a specific requirement for coordination of activities to assure a plan is in place so that workers are not placed in a position of danger, often referred to as the zone of danger, circle of safety, or the red zone.

I have had several cases on the railroad where they are "pushing" or moving the railcars, and they run over a railway worker who never even knew the railcar or locomotive was coming. It turns out that they don't use satellites or modern communication systems to keep track of each worker exposed to moving railcars. They have an outdated method of looking out for one another that fails again and again. I can't understand why they wouldn't incorporate more modern systems to protect their lives. I recently had a case where a rail worker lost an arm and a leg—a tragic and preventable case.

Workers are often caught in machines. According to the standards, it should be impossible to get into a machine because of guarding or lockout. This does not affect the human desire to get something done in a hurry and the pressures for production. I have had at least two cases involving trash compactors where the workers were having trouble getting the trash to flow through the machine, and they stick their arm into the compactor or even climb up on the compactor to get the trash moving. Usually, there is an electric eye that, at first, is not functioning when the trash is clogged, but then suddenly, that electric eye will start up, and the worker will die or be severely maimed. I will never forget that I once had a machine guarding instructor at the OSHA National Training Institute who lost part of his fingers. During the Christmas holidays, he was getting the snow off his driveway with a snowblower when it suddenly jammed up. He tried to unclog the snowblower by pulling out the snow without turning off the snowblower. Next thing you know, he lost the tips of several of his fingers. If the machine guarding instructors are losing their fingers in a machine, imagine how hard it's going to be for the rest of us.

Then there are the all too frequent electrocutions. Usually, this involves a piece of elevated equipment contacting the power lines, but it can also be the use of nonstandard equipment or people interacting with the electrical equipment who don't know what they are doing. I had a case where they were pouring concrete for a new gas station, and there are always power lines above the projects. A piece of equipment called a bull float was being used. This bull float is dangerous in that it has a long aluminum handle to allow the slab to be leveled and finished without stepping on and ruining the wet concrete. You can buy a bull float made of Fiberglas, which will not conduct electricity, but it is more expensive, heavier, and harder to control. This handle is usually long enough to hit the power lines, and sure enough, this happened, and a man was fatally injured when the electricity passed across his heart. The defense is that the power lines are right there, open, and obvious, and the worker should see them. The worker is working, and his focus is on his duties. It is impossible for that worker to concentrate on his finishing duties and watch the power lines at the same time. The most reasonable safety action in cases like this is to turn off the power lines and blockade or barricade any area where the bull float pole could contact the power lines. Other options include buying and using the Fiberglas bull float, putting a presence-sensing alarm on the pole, or using a small camera. Unfortunately, these companies just keep doing the work the old way, never making use of new technology, and these workers get maimed and killed.

Today, I got a call from a friend. He works at a large car dealership, and yesterday, which was Saturday, he tried to get into work past a doorway that had not to be cleared of ice and snow. When he stepped into the store, there was no matting inside the doorway, and a thin layer of ice had collected inside the doorway. He slipped and fell backward, and he hit his head on the doorframe and on the floor. He has a concussion, and we will see in a few days if this is a brain injury. I tell this story because this is a typical day for the safetyman. I found myself asking him who was responsible for clearing ice and snow, who was responsible for matting inside the doorway, who was doing the inspections, and if they were being done by competent persons. If he has a serious injury, it will be a big case, and his wife and daughter are going to need help. There is no rationale for a large company today not having a winter weather plan to prevent slips and falls. How are they going to make certain it won't happen again?

DISNEY AND SIX FLAGS EXPERIENCE

I think Disney and Six Flags have great safety programs and great safety people, but they are very large and diverse with so many potential hazards, and I have had a few adventures. The first was when I was with OSHA. I found out about an incident where Mickey Mouse fell from the geodesic dome at the top of Epcot. It was a simple matter of no fall protection and was quickly remedied, and I was told Mickey Mouse was fine. They just didn't like the visual of Mikey being tethered to the dome.

More interesting was the case I had at Disneyland in California. A woman was seriously injured in a boat on a water ride. Her boat went down the big hill and stayed at the bottom of the hill until another boat dropped right on top of her boat. It was great to be inside the back lot of Disneyland, and the people representing Disney were professional, friendly, and courteous. They gave us full access to the park and the water ride. I did some measurement and took some pictures, and I was later told that the lawsuit was withdrawn. I think they settled the case in a hurry.

Another time, I went to Six Flags outside Chicago, and I happened to take my nephew with me to take pictures, as he was in town and was an aspiring photographer and safetyman. We got up close and personal with the Superman Ride. This ride turns the passengers upside down, and a woman on the ground got hit in the head by an object that had fallen out of a passenger's pocket. My nephew was so helpful when he noticed there were coins in many locations under the ride that had fallen out of passenger's pockets, and there was no netting protection in those areas. This case was also settled soon after that investigation. Amusement rides are some of the most dangerous activities where families have exposure to a significant danger.

RED ZONE

When thinking about the dangers we all face in this world, the red zone concept is one that is extremely important and often overlooked.

It is not possible to protect every person or every piece of equipment with guarding or procedures. This is especially true with large pieces of equipment like cranes, boring machines, and, yes, roller coasters. There is always the possibility that a piece of the equipment might break or fall off. I had a case where a stowed jib fell off a fifty-five-ton rough-terrain crane. It turned out that this jib had fallen off the crane previously in some places around the world. According to some metallurgists, the vibration from the crane resulted in U-bolts that secured the jib to the crane in the stored position to be stressed and, eventually, to break. One solution to this problem is to remove the jib from the crane because it wasn't being used in any of the cases. Another solution to this problem, and many other safety problems, is to create a red zone to keep people out of areas where something could go wrong with a component of the equipment. This is frowned upon by many companies because it can slow down production, because work will be held up while the equipment is moving.

No red zone was provided in this case, and the man who was injured was checking the alignment of some piles during the pile-driving activity when the two-thousand-pound jib came down and hit him. Had there been a red zone, there would have been barricades preventing him from going into the danger areas while the crane was operating, and it would have prevented this horrible injury. There were also claims that there was an engineering defect on the crane that caused this jib to fall repeatedly, which was not proven to the satisfaction of the jury. Still, in my opinion, the crane company never fully investigated the engineering defect or took enough corrective action to recall the crane or warn the users. There were several witnesses that ignored well-known and established safety rules that could have prevented this injury. This case made me think that production pressures and the desire to make profits continue to take precedence over human safety.

There should be both an engineering solution and a red zone established. The companies involved in this case knew for a least sixteen years that there were problems with these U-bolts, yet they failed to fully engineer the solution or provide a required red zone.

It is common in industry and construction to hear verbal warning. "Don't move," "Don't back up," and "Watch for the flying, molten hot steel." "Don't walk under the crane." "Don't fall from the ladder, the scaffold, or the crane." Really these verbal warning just tell the story of a serious hazard that has not been corrected. One time, I was standing on the edge of an excavation, and it caved in just a few feet away from me and crushed a wooden ladder like it was a toothpick. Two people died under a car-crushing machine in a pit under a conveyor. Safetyman must go in to investigate. How can he prepare? What should he wear? The safetyman is going toward the danger. Wherever people have been hurt, he is there. I was with my friend and mentor, Shut 'Em Down, when we saw that an investigator had walked backward into an opening to accommodate a ladder. It is rare in construction that you see ladderway openings with gates or covers. We were both reminded that as inspectors, our attention is often diverted toward our inspection activities.

All you need is a safetyman with a camera diverting his attention, and if he backs up to get the picture, he can fall into those openings. The perfect scenario for a devastating injury. I sometimes ask the management of construction sites to take some action to protect my safety. This can include covering open holes because I might be distracted during the investigation or when taking photographs. Before I go anywhere, I try to get the lay of the land. I ask a lot of questions, and if I don't get good answers, then I proceed with great caution. Sometimes I just leave.

Every caper is similar in that it is me, the safetyman, against the world that doesn't care about safety and health. It's like people with grocery carts. It's not that much trouble for me to return my grocery cart to the cart corral, but everywhere I go, I see people are too lazy to do this, and when I park, I can see a grocery cart on a slight hill, waiting for the slightest movement or wind to send it into my car. You would think that people would just take a second to look after the interests of other people, but you could be wrong. That's why every activity is a caper for the safetyman. It's the idea that someone else has taken the time to look at safety and health for other people.

Even in offices, I see inconsiderate people stacking boxes in a haphazard manner that will eventually fall on somebody's head. I remember a case in an OSHA office where one of those big standards binders fell from a shelf and hit a compliance officer in the head, and he ended up with a subdural hematoma and brain surgery. The trouble is that we don't always anticipate what could happen. As we age and become less dexterous, the odds of an injury increase.

Going toward the road to push my garbage bin near the street is a serious hazard. The cars go at least fifty miles per hour in the thirty-mile-per-hour zone, and I have seen people looking at their cell phones swerve close to me when I approach the street. Two times in the twenty years, my mailbox has been hit and destroyed. This is just in my own driveway. Everyone should know that other people are often distracted, selfish, and not being safe. No matter who you are, you need to be a safetyman or a safety woman if you want to keep yourself and your family safe. You need to treat all your movements like a caper. If you don't plan and organize your safety and health, you could become a victim.

Nothing is more important in your life than planning and organizing your activities for the safetyman caper. I teach people about impetuousness as a real human problem. You want something done, and you want to do it right away. Move a tree or a rock, move furniture, cut a piece of wood, use a hedge trimmer, or even use a chain saw. But just doing it might be costly if you don't take the time to plan and organize your activity. Things have changed, and equipment and activities are far more dangerous than they once were, and we are more aware of dangerous injuries and death because of the media. When we were younger, nobody wore a helmet skiing or bicycle riding. Now they are looking into helmets for headers in soccer. I just learned that professional soccer players have as many as fifty headers per game. Nobody used to wear elbow pads or knee pads. Today, we are more aware of accidents. We think about avoiding injuries out of love for our own family. I believe that sending your children skiing or bicycle riding without proper protective equipment is negligent. I can remember when I first heard about Sonny Bono hitting his head while skiing at Heavenly Mountain in Stateline, Nevada. Michael Kennedy died in Aspen, Colorado, after hitting his head while playing football on the slopes. Neither were wearing helmets, and I remembered that I had skied for years without a helmet. It's slippery on those ski hills. Seat belts are no different and save so many lives. But school buses still don't require seat belts, and motorcycles can be ridden in many states without helmets. Remember that people are drinking and taking drugs, they are in bad moods, and they are late for work and appointments. We must prepare for the danger that we know is out there.

Preparation in safety and health for our caper starts with a simple concept called "hazard analysis." When I say these words, many people think it is a bunch of trouble or just words. It is simple. You can take a piece of paper and divide it into three sections, and on the left, write down the tasks you are going to attempt. Take skiing as an example. The task will be sliding down the mountain without injury. Next, in the center column, we write what could possibly happen.

While skiing, we could fall, we could hit a tree or another object, we could hit another person, we could fall off the lift, and we could have an equipment failure. That leads us to the third column, where we put what we should do to protect ourselves and our family:

1. Make certain we are fit and make sure we are acclimated to the altitude.
2. If you are not skilled at skiing, hire an instructor or take a lesson.
3. Learn how to safely use the chairlifts.
4. Make sure you have good equipment and protection—this includes the right-sized skis, helmet, goggles, and gloves.
5. Always ski at a controlled pace and never get out of control.
6. Avoid areas too congested with skiers.

Just a simple plan to avoid a serious injury or a ruined vacation. The question about these capers is how many people go forward with no plan at all. The idea of capers is to get people to realize that the world is not safe, and it is not going to take care of itself or anyone else. Going out into the world without a plan for safety and health could ruin your life. So, the next time you are going to do something that is unfamiliar or that could expose you and your family to danger, why not do a little planning?

LASERS

I was always surprised when I worked for OSHA how many companies used lasers. I am not talking about the ones we use for pointers in the classroom but big-time lasers. Those type B lasers. Pointers are dangerous enough, but the ones they are using in the industry to make straight lines in surveying and cut steel are amazingly powerful. Lasers are used to control the filling height for containers. A laser tells the bottling company precisely when the liquid has reached the desired level in the bottle of soda, bleach, and thousands of other products. A laser is defined as a device that emits light through a process of optical amplification based on the stimulated emission of electromagnetic radiation. There are thousands of different types of lasers, including gas lasers, chemical lasers, dye lasers, metal vapor lasers, solid-state lasers, and semiconductor lasers. Each one of these lasers presents special kinds of hazards and dangers to workers and the public. The danger we have all become more knowledgeable about is the laser that can be pointed into the eyes of an aircraft pilot, but there are lasers that will cut a human in two and lasers that give off deadly toxic gases. When I went to an establishment and found out that lasers were being used, I knew that I needed to do my homework. Like much of the homework of the safety and health professional, I needed to get the booklet that describes the laser and the hazards associated with that laser. There are questions about how the laser is being used, who is exposed to the danger, and how maintenance and repair are being conducted. I think we are going to learn more about lasers and the hazards of laser in the future as they become more and more prominent in our world. Thanks to OSHA, I had radiation training that included lasers at Northwestern University, and it opened my eyes to problems with controlling exposure to thousands of different lasers.

TRAINING ADVENTURES

I used to teach and train every week. That's where I learned a lot of this stuff. Standing in front of a room full of safety and health professionals is an adventure. For one thing, teaching adults can be difficult and problematic. Adults just don't learn the same as children. When you teach the same group for a week or even for two weeks, you get to know these people very well. I can tell you I made a lot of friends teaching all those thousands of courses, and the students in those classes never forget you when you are the teacher. This has been a little scary in later life when I go somewhere, and somebody remembers me teaching them a long time ago. Usually, they remember me well, and I don't remember them at all because I had so many students. They tell me I was entertaining. I always ask if they learned anything in those classes. There are always some good students who soak up everything. I have had students who knew more about my subject than me, and I learned from them. One time, I had to teach trigonometry, and thank goodness I had some students who knew it. Some students didn't like my techniques, and some didn't like me personally. I was frustrated when I would work so hard to teach the subject and entertain the students, and my efforts were not appreciated. The problem was that I had other duties besides the teaching. I felt like I was completely in charge of their lives. I was responsible for their hotel and their meals, and I had to deal with family emergencies, and then after giving them everything I had to give for so many hours, I would read an evaluation where some students thought I was average or missed some point they expected to be covered.

In every class, I got at least one evaluation that had a serious criticism. I often wished I was a celebrity performer; people might hate your work, but at least you don't hear about it every Friday on evaluation day. I used to complain to my boss at the OSHA Training Institute about the evaluations, and he would say to me that if one person said something critical on an evaluation, at least three other people felt the same way. When I first started working at the OSHA Training Institute, I had one course with terrible evaluations, and I am sure I deserved them all because I hadn't really developed as an instructor at that time. After I read all the bad news, I took the evaluation into my boss's office and told him how distraught I was about all the critiques. He read the evaluations and looked up at me and said, "If you keep this up, I will have to promote you into management!" I will never forget what he said that day, and it gave me the confidence to keep teaching and get better at my craft. I don't teach much anymore, and I don't miss it either. I do a few keynote speeches, which is easier. The adventure of being a safety and health instructor is a hard road, but it made me who I am to this day, and I don't regret a minute of it.

TEACHING OSHA IN IDAHO

I had a strange experience in Sand Point, Idaho. I went there to teach a class near the Canadian border. The class was filled with students who didn't seem to want to be there and didn't seem to like me too much. I must admit I was a little scared because the students didn't seem to like OSHA, me, or any outsiders. Living at the top of Idaho is tough, and the people in that class were living a hard life. I could see that they had a distrust of rules and regulations. They were taking the OSHA class just so they could get an OSHA card and get more work. At that time, much of the construction industry required their workers to have an OSHA ten-hour card before they could work. There was demand for the OSHA cards. Things got so bad with the need for OSHA cards that there was a scandal in New York. Some outfit was selling the OSHA cards without providing the training. Some students told me that they could buy OSHA cards at other locations. After this scandal, OSHA started numbering and tracking the real cards. I can tell you that these Idaho guys would have preferred to buy a card rather than spend ten hours with me going over the standards.

Safety training can be frustrating for everybody when the students do not pay attention or learn what you try to teach them. There were a couple of times when I refused to issue cards because the class or individual students were so inattentive. Students must pass a test at the end of the class, but it's a basic test and doesn't really indicate how much learning has occurred. The class in Idaho gave me some insight into safety training. If adult students are forced to attend training and are not interested in learning about safety, it can be a terrible and wasteful situation. I was glad to leave Idaho, but I did the best I could to try to motivate those people, and I know they learned about some hazards. Even this class enjoyed the demonstrations I conducted about explosions and soil mechanics. It's important to note that most of the places where I conducted safety training, the students were eager to learn and appreciated this training. They realized that this training would lead to higher salaries and more career opportunities.

TEACHING AT MOUNT RUSHMORE

There are many sweet assignments from the OSHA National Training Institute—Hawaii, Puerto Rico, and the possibility of anywhere in the world. Sometimes a place that seems mundane turns out interesting. I was sent to South Dakota for a class for the Park Service. I had never been to South Dakota, and it wasn't long before I found out it was home to Mount Rushmore. On the first day of class, I learned that my students were running the park, and they told me stories about Mount Rushmore. I was especially interested in a secret room and tunnel behind Lincoln's head. It was a fun class, and on Wednesday night, they promised to take me to the secret room. It was a pretty good climb, but to go where tourists never get to go was an experience I will never forget. I was impressed with their safety program and the conditions I encountered while making the climb. Mount Rushmore is an amazing place, and everyone should see it during the day and at night. I learned that terrorism was an important consideration at the monument, as it is at all monuments, and I was impressed with the readiness of these men and women who were my students. On Thursday night, I decided to go out to the monument one more time before I left the park the next morning. I drove my rental car, and I had no idea the speed limit in front of the monument was fifteen miles per hour. I was speeding at about twenty-five miles per hour when the unmistakable flashing light of a police car came up behind me. I pulled over and waited for the police to come to my car. The officer came to my window, and it was one of my students with whom I had become friendly during the week in class. He smiled at me and told me to slow down.

WORKING IN SALT LAKE CITY, UTAH

I had been to Denver hundreds of times but never to neighboring Salt Lake City, Utah. I went there for an accident investigation that was postponed because one of the attorneys got sick. I had a free day in Salt Lake City and decided to check out the Salt Lake City Church at Temple Square. This turned out to be an amazing experience. I am not sure if they saw me on their cameras, but two young women and a middle-aged man came to me and asked me if I wanted to tour the facility. I couldn't believe my good fortune. They took me all around the campus. I was not allowed inside the church chapel, but everywhere else I went, the place was enchanted. I went into a hall where the 360-member Mormon Tabernacle Choir sings to three thousand people, and it was impressive. The place was built in 1867 and has a roof that is 150 feet across and 250 feet long. They also took me into a private area with portraits of all the leaders. I was treated almost like a member of the family, and I learned a lot about genealogy, which I found out is a great interest of the Mormon people. I let them know that I was a safetyman, and I would be looking for hazards on the sprawling thirty-six acres. I have never been any place where I could not find a single violation, except this wonderful place.

BUCKEYES AND CHILI

For twelve years, my company had a contract to teach construction safety at the Bureau of Workers' Compensation for the state of Ohio. This contract was not easy to get. Every year, we had to fly into Columbus, Ohio, and compete with other training vendors and be evaluated by a panel of administrators. It was like *American Idol* for training. It was great to win those contracts for all those years, and it offered the opportunity to travel all over the great state of Ohio to teach the construction safety course. We taught in Columbus, Middleton, Akron, Cleveland, Cincinnati, Canton, and even places like Coshocton. I remember being in Coshocton two times. It was a beautiful place on the Walhonding and Tuscarawas Rivers with a wonderful old hotel and restaurant. When we taught a course in Canton, I had the opportunity to visit the NFL Football Hall of Fame. The best memory I have of Ohio has to do with the Skyline chili that they have all over the state. I am not sure what they put in that chili, but I can tell you it is delicious, and we would eat there almost every day. I still like to buy it frozen and eat it at home, but it is not the same. I loved it so much that one time we were in the parlor—that's what they call them, chili parlors. (The words *chili parlors* remind me of the Elvis Presley song "Are You Lonesome Tonight.") Anyway, they have a Skyline neon clock on the wall, and I asked the manager how I could get one. To my surprise, he told me he had an extra one, and I could buy it at his cost. Below is a picture of that clock that I still have to this day.

TOPICS

SLIPS AND FALLS

MOST INJURIES ARE A RESULT OF SLIPS, TRIPS, AND FALLS.

When humans started to walk on two legs, instead of four, they were not as close to the ground. The average adult's eyes are at least five feet from the ground, and it is not possible for us to walk while looking at the ground. Humans have eyes in front of their heads, near the top of their heads. The possibility of falling becomes evident. The problem I have noticed with slips, trips, and falls is that when we are children and close to the ground, we slip, trip, and fall constantly and almost never get hurt. It is like a rite of passage. There is some point in every life where a person reaches an age or a condition where a slip, trip, and fall is not a good thing at all. We all know this, and it seems simple, but it doesn't change the fact that the world is full of these trips, slips, and falls, and we somehow think it is not a serious problem. There are two considerations about falling that most people don't really appreciate. One is that as you get older, the bones and body parts become more fragile and brittle, and the opportunity for injury is greater. The second consideration that's not fully appreciated is humans might fall into something that is sharp or will penetrate the body.

I had a sad case in Alaska where a young child was getting off the school bus, saw his mother, started running toward her, and tripped. His eye contacted a stick from a tree root coming out of the ground that penetrated his brain and caused a serious brain injury. Most of us don't even think about sticks coming up from the ground, but I have had other cases where people had their bodies and vital organs impaled by sticks, wire, rebar, and other objects. Ski poles are notorious for doing damage. When you look at all the cases and the cost of these slips and falls, a safetyman gets the impression we are not doing enough.

A recent case involved a woman who tripped and fell on the remnants or metal stub of a removed parking meter. A piece of metal was sticking up in the middle of the sidewalk. She tripped on the stub and fell on top of the sharp portion, and it changed her life forever. City contractors who do the job of removing parking meters must make certain that any tripping hazard is blocked or barricaded until the sidewalk can be repaired. Too often, responsible parties take it for granted that people can see tripping or slipping hazards, and they think that even when people fall, they will not be seriously injured. Assuring that means of egress and access are safe and unimpeded is essential for safety.

PERSONAL PROTECTIVE EQUIPMENT

It is far better to eliminate the hazard completely, like guarding the machinery, but some machinery, like grinding wheels, can't be fully guarded, and we are left with personal protective equipment being an essential option. I remember the first time I appreciated the importance of personal protective equipment was in the steel mill. I was with a crew of workers whose job was to tap the blast furnace. To do this job, you must go right up to the furnace that is melting iron ore, and you must release the molten metal. Personally, I would like to see a robot do this job, but I can tell you, without the provided protective clothing, we could all have died that day from the heat. Many companies think they can just hand out personal protective equipment and workers will use it, but my experience is that the workers do not want to wear extra stuff, and if you give them the wrong stuff or they use it the wrong way, you can actually create more danger.

Many construction companies I have seen hand out lanyards and body harnesses and think they are done with fall protection. The problem with this is that those lanyards and harnesses won't do any good if they are not attached to a proper anchorage and the harnesses are not properly adjusted. There must be a system of planning for fall protection. A company can be setting a trap for the worker by giving them a false sense of security when they actually have zero protection. It is normal human behavior for workers to resist or even forget to use their personal protective equipment. There is also a likelihood that they may not use it the way it was intended.

This disparity between human behavior and the need for protection is a great problem for the safety professional. The workers need to understand the need for the equipment, be trained on the proper use of the equipment, and be provided equipment that is compatible with their bodies and comfortable for their use. Most people reading this section will tell me that they have used a hammer and nail without safety glasses, and I have done it myself without thinking. Fortunately for me, I always have safety glasses on my face but not with side shields. Those side shields are important. I have had some close calls with branches coming around my glasses in the backyard. The trouble is if a piece of metal or a twig flies into our pupil, we could be blinded for life, and at that point, we are going to feel stupid for not protecting our eyes. I have met hundreds of people who have lost their eyes, and believe me, if they could go back and do it differently, they would wear the safety glasses and the side shields. This does not change the fact that people are still not doing it. We can think that people will do it, but if they are not doing it, we are only fooling ourselves.

Hard hats are uncomfortable. I don't like wearing them myself, but it is so much better than hitting my head. Preventing a concussion or a brain hemorrhage makes me suffer from hard hats as much as I can. At this point in my life, I will not even go into my basement without a hard hat. There are some low pipes, and I hit them even with the hard hat, and it doesn't hurt at all. I think hard hats can be more comfortable, and they should always have chin straps so they don't fall off. I have worn hard hats that hurt my head all day, and some others that I hardly even noticed. There are too many types of personal protective equipment to go over them all here, but let it be said that we have got to find a way to make people want to use it and want to protect themselves with it.

SNOW AND ICE

The earth may be warming, but snow and ice are still a problem in most places. There are several methods available to deal with snow and ice to protect humans from falling and hitting their heads or other body parts. The first is shoveling snow and chipping ice. Unfortunately, this process that is helpful for cars and parking doesn't solve the slip-and-fall hazard. For the slip-and-fall hazards, we have salt and other chemicals to melt the snow and ice quickly. This salt, unfortunately, has environmental side effects, and in some places like

Colorado, they prefer the use of sand, which doesn't melt the snow and ice but increases the friction. The main idea is that we need to do whatever is necessary to prevent these slips and falls that are so costly to society. Not enough attention is given to footwear. Spiked footwear is extremely effective against slips and falls on snow and ice, but most workers and people wear sneakers that offer little protection. Once again, the human is their own worst enemy when they don't take action to protect themselves. As a safetyman, I believe we need to make better use of helmets and hard hats on a regular basis when there is snow and ice. The biggest problem is hitting the head, and if we are not able to change human behavior and exposure to the hazard, we can at least protect from a serious injury.

I had my very own experience with ice about ten years ago when my family was going on vacation. It was early morning before sunrise and dark outside, and I had successfully driven to the airport on icy roads. We were late for the flight because of those icy roads, and after parking at O'Hare, we were rushing to make the flight. Everything was fine until we hit a sloping sidewalk that led to the tram that was coated with a fine layer of what is known as black ice. To this day, my daughter can't believe I survived. I hit that ramp, and she says my legs went higher than my head, and I flew in the air. She thought I had hit my head, and if I had hit my head, surely, I would be dead or, worse, with a closed-head injury. By some miracle, I didn't hit my head, but my ribs hit the suitcase, and we found out after we came back from the vacation I had broken two of my ribs. I was in so much pain I couldn't breathe, and the vacation was completely ruined. From that experience, I now have more sympathy for victims of ice and snow accidents. I also realize the importance of protecting people who are not expecting to encounter black ice. If someone at the airport had just put some salt on that ramp, I would have enjoyed my vacation with the family.

THE SLIPOMETER OR TRIBOMETER OR WHATEVER

If you are a safety professional and you are out there being a superhero, you are going to find that a lot of people slip and get hurt on floors and other walking and working surfaces. There is a problem with ice, oil, and water. I have seen fruit, flowers, and even urine and feces. The question is about the slip resistance and friction coefficient of the flooring materials. This also relates to the type of material used on the floor, whether floor matting is used and many other factors. There are important safety issues involved in determining whether the floor meets safety standards and determinations of relative slipperiness.

As a scientist safetyman, it is necessary for me to have an instrument to help me determine compliance. I was working with the American National Standards Institute (ANSI), so I knew there was a standard of 0.5 friction coefficient, and with an instrument, I could test compliance or noncompliance. I got a robot called the Tortis slipometer or tribometer. I must say that I love this instrument because it is completely objective. I can plug it in and run it across dry or wet surfaces, and it will give me a reading of friction. Then the only question is whether it is the right reading or not. I now bring a backup handheld friction instrument as a control to show that the two instruments are in the same ballpark. I also invite the other safety expert to bring their own instruments for comparison purposes. This is a powerful tool in the evaluation of slip-and-fall cases, and I think this is the way it should be for all safety professionals. If a technology exists to measure a physical property, it must be used. There can still be many questions in a slip-and-fall case about matting, locations, and human factors, but having physical evidence is powerful. I also wonder why manufacturers, architects, and owners aren't conducting this testing before they decide what materials to utilize. I have seen a smooth tile used next to swimming pools and hot tubs where much safer friction tile is available for the same cost. I have been told that they chose the smooth tile for aesthetic reasons. It seems to me that retail establishments that invite patrons into their businesses should know everything about walking surfaces for the protection of their employees, customers, and their bottom line. I am surprised that one of the shops or areas within a giant store has carpet or friction tile in some locations, but in other areas subject to water accumulation, they have smooth tiles with little friction protection.

FALLING AND FALL PROTECTION

Falling is probably the most common cause of injury. Most people have not fallen from heights, so they think they won't fall. The falling issue is complicated with compromise. The standards allow no fall protection on ladders even though there are ladder-climbing devices that solve this problem for fixed ladders. There are so many exceptions, including exemptions for residential housing, leading edge work, and even allowing safety monitors in lieu of actual fall protection. When teaching this subject, a warning was given to students that, in the end, they would find the fall-protection regulations, customs, and practices complicated and confusing. Many safetymen find we need to be creative to gain safety and health for the people who will be counting on us to return home to their families each day. The matter of fall protection is further complicated by language, such as "conventional fall protection" and other kinds of fall protection that are not really fall protection at all. See, it is confusing.

To make matters worse, there is a question regarding which employer is responsible for fall protection. To me, the question is about which employer is responsible for the fall-protection plan. Any employer, and especially the employer of the employee, can provide lanyards and harnesses, but not every employer has the ability or responsibility to create the fall-protection plan. This plan must be in place to allow the lanyards and harnesses to be used in a safe and practical manner. People need to be tied off, and the capacity of the anchorage is required to support, at a minimum, five thousand pounds per person. The safety professional must know the engineer for the anchorage or at least the engineering plan and information to determine the feasibility of the personal fall-protection equipment and the five-thousand-pound anchorage. The employer responsible for this type of engineering and planning is likely to be a controlling employer, such as a general contractor or a prime contractor in a situation for rack storage. It could even be a specialty contractor, such as an equipment contractor, if, for example, it is determined that workers will be tied off to the equipment, such as a crane of a forklift. See, this is complicated. Many times, the answer to providing proper and safe fall protection is following the safety hierarchy that dictates the need for an engineering assessment. At that time, systems and equipment will be determined to make it impossible for workers to fall, or if they do fall, they will be caught by shock-absorbing lanyards or even personnel nets. Also, be aware that there is a big difference between personnel nets and debris nets.

NIP POINTS, PINCH POINTS, AND IN THE BITE

These are all names for the same thing. Think of a doorjamb but way more dangerous. All machinery has areas where your body or your clothing can be smashed or caught, and these injuries are terrible. I have seen the drill bit and the glove injury too many times. Whenever you have a motorized spinning part of equipment or machinery, you are just one move away from losing your finger or your hair or even another portion of your body if clothing or hair gets caught in the rotating movement. Rotating parts are just one kind of danger; you also have the danger of any moving part, such as a conveyor, where there is an opening and where any portion of your body can contact the moving parts. Shafts, pulleys, belts, and all moving parts can be a serious hazard. It seems obvious, but think of an unguarded fan; when the blades are moving fast, you can't see them. I can remember when fan blades were not well protected. Today we know and understand that people can forget or make a movement without understanding possible contact with a dangerous location, and they will end up with a significant injury.

Workers' and peoples' clothing have been pulled into rotating parts of drills and augers. When this occurs, they almost always blame the workers or the people. They will say people can see the danger, and if they get into the danger, it is their own darn fault. I have a case where a worker was operating an auger, and he needed to clean it off. It is much easier to clean it while it is still running. His coat got caught in the turning auger, and it pulled him in, and he suffered and eventually died. The truth is that any of us could be wearing gloves, a coat, a sweater, or even a tie, and when anything gets pulled into these rotating or moving parts, it ends up with suffering, permanent injury, and even death. This is from just one mistake, and those rotating parts are required to be guarded.

I had one case at a steel mill where a worker's safety lanyard got pulled into an auger and dragged him into the machine. It is hard to imagine the horror of such an event. My experience tells me that people, in general, have no idea of the seriousness of this hazard. Even a battery-operated portable drill can take you down if it grabs your glove or your tie. You might not know that there are guards for these drills. There is an inexpensive guard for drills and augers that covers the rotating portion of the bit and then moves up the shafts as the drill or auger digs into the ground or the material. This means the hazard is eliminated. The problem is that nobody is using these guards, and they are not even that easy to find in stores. When a consumer buys a drill, they are not likely to be a drill or safety expert. There should not only be an offer to sell the guard, but there should be a warning about the rotating hazard. How many readers think they can wear gloves while using a drill press? It almost seems logical to protect your hands from splinters from the wood or sharp metal you might be drilling. *Please* don't wear gloves, because it will destroy your hand when that glove gets caught in the rotating drill bit.

MACHINE GUARDING

The word guard is defined as it shall prevent, under all conditions, a person from encountering moving components (1949 Model Code of ILO). Any company that knowingly allows their workers to be able to have contact with dangerous machine parts has violated guarding requirements.

It seems obvious to say the machine that exposes the worker to danger must be fully protected. Today we have the equipment and technology to make it a reality. There is no excuse for a fatality or mutilation in a machine or any moving part of a machine or equipment.

FULLY GUARDED

When a machine is fully guarded, there is no danger unless the guards are removed or missing. The guards are required to be adjusted to protect the smallest body part, usually the pinky finger, and the workers may be there only to watch or feed material into the machine through a tiny opening. Of course, when lubrication, maintenance, or adjustment are necessary, or when the dies need to be changed, the guards are going to be removed, and that is when the lockout/tag-out standard takes the place of the physical guarding. Before those guards are removed, that machine must be determined to have zero energy and to be completely safe by a competent and qualified person. Lockout/tag-out is an elaborate and time-consuming procedure, but it is also a life- and limb-saving procedure. Before a worker puts his or her hands in any danger zone, the machine is required to be tested to determine that it cannot be activated. The lockout/tag-out procedure is often avoided by companies to try to keep production running, and sometimes shortcuts are used to try to keep the line moving. These are situations where a company can run into trouble because an energized machine and workers in the bite are an illegal and dangerous condition that too often results in a horrible mutilation or death. If the machine is energized or there is any type of energy in the machine, then no person should be able to get near any danger. The danger can be the point of operation or any other moving part. It can be a pinch point or rotating, moving, or reciprocating portion of the machine or equipment. Common dangers include live electrical energy, pinch points such as on power presses, rotating parts, such as a turret lathe or shafts, and V-belts and chain and sprockets. No moving part should be energized and unprotected. Physical guarding is one of the most fundamental and effective protections for human life and safety, if those guards are in place.

OTHER DEVICES

Sometimes there is a need to interact with the moving parts of a machine without a physical barrier guard in place. In these cases, there are other types of protection that are sometimes utilized. Two-hand tripping devices or two-hand control devices are an example. There are also light curtains and presence-sensing devices. These devices keep the hands of the operator busy and away from the danger or turn the machine off when approaching the danger. The idea for these devices is that the machine can only be actuated when the hands are an appropriate distance away from the bite (safe distance), or the machine will turn off when a portion of the body gets into the bite. These devices are *not* as effective as full-barrier guards or lockout/tag-out procedures. Although allowed in some cases, these devices should be avoided whenever possible, as the safe distance only protects the operator, and every presence-sensing device can fail. Another problem with these devices is when the machine is moved, the hazards could also be moved and require additional guarding or safeguards to prevent contract with the danger zone of the machine.

LIGHT CURTAINS AND PRESENCE-SENSING DEVICES

There could be an invisible shield protecting the danger that immediately turns off the machine. If you break a force field or the machine senses the movement, warmth, or temperature of your body, it turns off the machine and stops it before an injury takes place. This could solve the problem of needing access to the danger, but the problem is that these devices might not work the way they are intended, or they might not get the maintenance they need to function perfectly. One failure can result in death or mutilation. To me, this is like people who stick their hands or legs into the elevator door to stop the elevator. It might hold the elevator, or it might just take your hand or leg down to the next level. I just saw a video where a bystander saved the life of a small dog whose owner walked into the elevator with the dog on a leash. Only problem was that she didn't make certain the dog was in the elevator. The bystander managed to get the collar off the dog before it was too late.

BRAKE MONITORS

As a result of concern that the above presence-sensing devices might fail over time, many manufacturers have installed brake monitors. When the force field is broken, the most important thing is for the machine or machine component to stop immediately before it can crush a worker or a body part. If the brakes wear down or don't stop fast enough, the result is a serious injury. Therefore, secondary protection is installed that will monitor the brake and stop the machine if there is any indication of wear or failure. These types of safeguards are known as redundant safeguards, and many have proved to be effective. My position is that all these other devices can fail, and as a safetyman, I want total protection in the form of a physical guard or lockout/tag-out procedures.

ROBOTIC PROTECTION

If, for some reason, physical guarding or lockout/tag-out protection is not available, I would look to automatic or robotic operation to keep people away from a danger zone. I discourage my clients from using any protection that requires extensive maintenance or that can fail without warning.

BARRIERS

Fences and gates are effective means of guarding dangerous areas until they are entered. Sometimes, fences and gates are interlocked with the machine. These interlocks are easily bypassed with duct tape, and workers can climb over fences or behind gated machines. I have even seen visitors and contractors go behind a machine to have a smoke and end up inside the machine. Barriers and gates are only effective if they are tested and proven to be safe under all circumstances. Too often, there is a belief that putting up a gate is all that needs to be done when an interlock is required by law. Think of it like your washer and dryer. They both have a door or gate, but they also have an interlock to turn off the machine when the door is opened.

THOUGHTS ABOUT GUARDING

The biggest problem I have seen with guarding in the field is companies using equipment that is old and not adaptable to modern guarding techniques and concepts. They often start out by telling me that they have these old machines, and as soon as I hear that comment, I know there are likely to be problems with machine guarding. As a rule, most of these old machines can be guarded effectively, but the cost is high, sometimes hundreds of thousands of dollars, to integrate automation into machinery that was designed during the industrial revolution. Using these older machines can be profitable, and there is reluctance to upgrading at a high cost until a terrible injury occurs. Being proactive means understanding that if an injury does occur, it could cost millions in fines, and they still must make the expensive upgrades. Usually, the thinking is that they can train their workers to be so careful that nothing will happen. I am all in for good worker training, but attempting to train workers to not make human mistakes is not a good or acceptable practice. It has been a battle of my life as a safetyman, to attempt to get companies to be proactive with their thinking and their money.

TOOLS

EVERYONE READING THIS HAS HAD THE EXPERIENCE WHERE they need a wrench or a screwdriver to fix something around the house or to put something together. Maybe you don't have the right tool, but it is too much trouble to go out to the store and spend some money to buy the right tool for the job. Attempting to use the tool that you happen to have around the house or some cheap tool you have purchased at the grocery store can result in a serious injury. Most people think nothing will happen if a wrench slips on a pipe or if a screwdriver slips or breaks, but I have seen broken knuckles and stitches so many times when the wrong or defective tool is used. My advice is to buy the new and proper tool or, even better, to get a professional to do the work. I recently had a friend try to repair their own air-conditioning unit in their home, and they are now in the hospital. They were trying to save some money and used the internet as a guide. The problem was they were not in the air-conditioning business and did not have the proper tools or skills to do the job.

STEEL MILLS

The steel mills have been a big part of the safetyman experience. I have found myself going to the mills repeatedly, where thousands of people are working hard with raw materials and moving them with heavy equipment to a place where they can be ignited to extreme temperatures and formed into materials that make our modern world what it is today. It is not surprising that there are many injuries and fatalities in steelmaking because of the nature of the work. The heat involved is extreme, as steel melts at 2500 degrees, is poured at more than 1500 degrees, and can take six weeks to cool to 1000 degrees when it solidifies. In addition, the mill is in a constant battle with the materials that are trying to contain the heat for the process. The molten steel is contained by refectory material that has a limited life span and is constantly in danger of failing and causing a catastrophe. It is amazing how well the molten metal is contained within a steel mill. Shutdown, contractors, and thousands of monitoring and testing devices are used, all ready for one failure or mistake to result in a serious injury, fatality, or catastrophe. A list of the dangers in the steel mills includes significant hazards at the blast furnace, where I have spent much of my time. Workers and safety people must deal with heat, noise, dust, liquid metal, slag, gas poisoning potential, moving equipment, locomotives, fires and explosions, and working at heights. There are road hazards, coke ovens, heat, dust, smoke, constantly moving equipment, noise, liquid metal and slag, and gas poisoning. There are other hazards in melting shops, rolling mills, and power plants. Even under the best of circumstances, there is always a potential for injuries and death in the steel mill. Upgrades in technology and the use of automated equipment and robots are removing the humans from the danger in steel mills, but there is room for more improvement. Nobody should be subjected to the possibility of exposure to molten metal, poisonous gasses, or uncontrolled explosions. When you want to understand the problems with assuring that workers are not subjected to the possibility of death or serious physical harm, which is the promise of OSHA, it is a good idea to go back to the steel mill and see that this promise is not guaranteed to the steelworkers near the blast furnace. A release of steam or poison gas, or failure of containment, still means certain death. Every time I visit the steel mills, I am reminded of the need and the cost of technology that will make this promise come true for every worker. Right now, we do not have intelligent robots to make changes and repairs, and human workers are risking their lives to make the steel we need for our modern world.

CONCRETE AND CEMENT

Concrete and cement are some of the most amazing things I have learned about in safety. If you look around the world, our society would not be possible if not for this mixture of calcined lime and clay, sand, stones, gravel, and water. The material starts off soft and can be molded into almost any shape, and then it cures to be hard and strong. It's pliable in that it can be formed into almost any shape, and it can be reinforced with steel and other materials. The curing chemistry is interesting. It takes around thirty days to cure in an exothermic reaction, and it keeps getting stronger and stronger over many years. Admixtures that are added to concrete and cement to give it different properties. Some of these admixtures make the concrete cure faster or at different temperatures. One of the most important safety aspects of both concrete and cement is that there must be sufficient waiting time for it to cure before it will sustain the anticipated load.

There are significant hazards associated with cement and concrete. The hazardous effects of the materials used in concrete and cement in construction work occur every time a bag is opened and people are exposed to the dangerous silica dust. We can get a significant exposure just patching a crack in the sidewalk or driveway. Before I got involved with being a safetyman, I had no idea how hazardous and dangerous the dust is to our lungs. I can remember as a child putting my initials in some soft curing concrete or cement before it was dry. I had no idea that these chemicals can burn your skin right off your bones. It wasn't until OSHA sent me to concrete school at the University of Wisconsin in Platteville, Wisconsin, that I learned that the same magical stuff that allows us to form buildings and structures is highly corrosive and dangerous.

Many accidents, deaths, and injuries have happened because of production pressures when there is not enough time given for the concrete to cure. There was the Ambiance Plaza incident in Bridgeport, Connecticut, where a sixteen-story residential project fell on April 23, 1987, killing twenty-eight construction workers. Concrete fell upon workers because of the lack of lateral stability. Then there was the Hyatt Regency walkway collapse on July 17, 1981, where the walkway collapsed onto a tea dance being held in the hotel's lobby, and the Willow Island disaster on April 27, 1978, where falling concrete caused scaffolding to collapse, killing fifty-one construction workers. Outside the USA, there have been more incidents, including the collapse of the Hotel New World in March 1966 in Singapore.

Concrete can fail to hold an anticipated load or an unanticipated load, or the pressure buildup from the chemical reaction within the concrete can cause forms to burst, and it will explode on the site, injuring the workers. All too common are cases within the construction industry where workers have been burned by the hazardous chemicals, including lime, within the concrete. The concrete or cement gets inside the boots of the workers, and they don't even notice it until after work when they take off their boots, and the skin falls off their bodies. This same phenomenon happens to people when they are doing a project at home with cement. They may not be warned about the hazard or read the label, but if you expose any area of your skin to the cement, you can do some serious damage. Please don't forget to wear a respirator when you open those bags, as there is a high percentage of silica dust that will be emitted.

CRANES AND RIGGING

Cranes are the workhorses of construction, and as a result, the safetyman has spent much time with crane standards and even some reporters. I was quoted in the *Wall Street Journal* when a crane fell in the streets of New York City. My quote was about the need to inspect these cranes, and the people who inspect the cranes must be qualified. That sounds simple enough, but the truth is many of these cranes are not inspected by qualified people, and since all of them are built like a seesaw, when the crane or forklift gets out of balance, things get ugly really fast. If you think of cranes and lift trucks as scales of balance, it is easier to understand their strength and fragility. Unlike a scale, there are many parts not easily visible to the inspector. Add in all the tension and stress, and something could be wrong. Even the load line, that thick cable that looks indestructible, can have serious damages from shock loads or side pulls that can't be seen from the outside. These cranes and forklifts must be examined by people who know what they are doing, but even with careful examination, there still can be unseen defects, issues, or problems, especially in older equipment. The industry treats older equipment just like the new equipment, probably because it is so expensive. The cranes, even under the best of circumstances, can fail or fall. High winds or sudden stress from an earth tremor or a thousand other factors can bring them down. This means that the best safety for a crane under any conditions, and especially under conditions of a working crane, is to keep the people out from under the crane. Establish the red zone, barricade the craning area, and minimize the exposure to human life.

There will be times when certain workers need to be near the crane but never under the load. They need to hook and unhook materials, change the rigging, and perform maintenance. Situations where workers are close to a working crane should be considered special situations with special precautions. There is a concept in the standards called the "critical lift" that requires scrutiny where there are exceptional loads or people near the working crane. The critical lift concept should include whenever humans are exposed to the crane movement or load. There are too many fatalities where workers are killed while interacting with the crane while it is working. This happens often with the large tower cranes, with all its moving parts, many that are used to raise or lower the crane. There must be strict rules and good communications between the operator of the crane and the workers. One mistake can result in the loss of a worker. The mechanism that raises and lowers the crane is called a climber. These climbers have been the cause of many fatalities. The operator of the crane or the chief of the crew might not see where the workers are located. There is no reason whatsoever cameras can't be used on cranes to save lives.

Another crane just fell on the streets of Seattle. There is only preliminary information, but they are already saying it was probably a human error. The machine does not fall on its own. The crane is in a delicate balance, and all the rigging, whether done on the crane or with another mechanism, like a hoist, is under extreme tension. A crane, a forklift, and a hoist can fail at any time, so controlling personnel in the area is essential. An old car has a better chance of breaking down than a new one, but for some reason, with this equipment, the attitude is that they are all the same. Theoretically, they could be the same if there was perfect maintenance and no hidden wear and tear, but the older equipment is not as safe or as easy to maintain. Every piece of equipment involved in material handling must meet standards that include proper maintenance, and they also require that people running it and in the area are never exposed to the load or the raised equipment. Most safety professionals know that hydraulic equipment of any kind can suddenly fail. There are elaborate safeguards for hydraulic lifts that are used in service garages and oil-change businesses, but those same safeguards are not normally found on the equipment in the field. Even with elaborate safeguards, people get hurt or killed when the lift loses fluid pressure and suddenly falls. Hydraulic equipment used with cranes, lulls, forklifts, and other material-handling equipment is just as susceptible to leaks and failures. The only protections and safeguards are inspection and maintenance. People make mistakes, and they are not always diligent in their inspections. It is a human quality to sometimes not be conscientious or to be distracted. Anyone can miss a small detail.

Rigging is a separate category. There are many kinds of rigging, including nylon, steel, cable, and chains. There are different sizes and capacities. There is a lifetime of learning about cranes and rigging alone. One of the biggest problems with rigging is there are hidden defects. It is well-known that cables might have internal damage that affects the structural integrity of the rigging that will never be seen by the naked eye and may only be found with sophisticated testing equipment, including x-ray. One of the biggest problems with rigging is that the typical workers without sophisticated equipment or any engineering training are the people who most likely will be inspecting the rigging and deciding that it is acceptable or not acceptable. Slings are a type of rigging that needs to be arranged to assure the load is balanced and stable. There are charts to assist with this task, and this is an area of concern when lifting enormous weights. It is rare for me to go to a construction site and not find at least one piece of defective rigging. I have also seen the result of failed rigging where the entire load fails and falls. This situation injures or kills workers and destroys the jobsite.

CONFINED SPACES AND PERMIT-REQUIRED CONFINED SPACES

Here is another area that can be a little confusing for the student. The standards are divided into two separate sections. It might have been better if they were not separated. The main thing is when you go into a place that you have never gone before, you better make certain (1) there is air down there and (2) there is nothing that is going to make you sick or kill you now or in the future. This makes perfect sense, not common sense, but it is surprising how many people, both in the workplace and outside, don't think they will ever enter an area where there is no air or where there are contaminants so toxic that they will shorten or end their life. I go to these places both before and after people have died, so these elements are important to me to keep on living.

We don't use canaries anymore for our protection. In the olden days, they had a cage with a canary in the coal mines, and they would keep an eye on that canary, and if they saw the canary had died or was in distress, they would evacuate the mine. Today, we have instruments and meters, also called sniffers. These instruments are great and cheap. I have seen them for as little as thirty-five dollars, and they will save your life as well as mine. Most meters will calculate oxygen, which is the most important thing, because even if there is toxic material down there, you will still need to breathe long enough to detect it or run from it. Most instruments will also calculate carbon monoxide and hydrogen sulfide. Carbon monoxide is well-known because of vehicle emissions, and it is a by-product of any combustion, including your charcoal barbecue. Less known is the danger of hydrogen sulfide, which at extremely low levels will take you off this planet. Most people don't know that hydrogen sulfide is naturally occurring and comes from the ground. I once had a case where three groundskeepers working in a sand trap on a golf course were overcome by hydrogen sulfide. There is a lot more to confined space entry besides testing with instruments. You will need an attendant and means of escape and a fully developed program if you intend to go where no person has gone before you.

A confined space is defined as a place where there is no human occupancy, but it is small enough to allow someone to get in and has the potential for not enough air or toxins that can injure or kill you. Enough air is considered to be 19.5 percent oxygen, and even though you might find 19.5 percent at the opening or in other locations, there might be a source within the confined space where there is 0 percent oxygen, so a person like me has to test at all levels and all locations if they want to see their grandchildren get married. Also, since normal oxygen in regular air is 21 percent, if you only have 19.5 percent, it would be pretty important to find out what has displaced the 1 1/2 percent oxygen, because if it is something like hydrogen sulfide, you will die. I intentionally did not get into the difference between confined spaces and permit-required confined spaces because I believe they all need permits. Basically, a permit is just a checklist to make sure you didn't forget anything. A pilot of a jet airplane is required to go through a checklist before each takeoff and landing because it involves so many lives. It is no different for a confined space.

ELECTRICAL

Electrical hazards are some of the worst. You can't see electricity, and when you start to understand voltage and amperage, you realize that electricity is complex and can be unpredictable. Electrical hazards include lightning, static electricity, and all the voltages and step transformers. Concerns abound around flammable and even combustible materials, and there are many different precautions, including ground fault circuit interrupter (GFCIs), grounding, and bonding. There is special training and personal protective equipment used only for electrical hazards, such as hot sticks and power line grounds. The biggest problem with electrical hazards is the people working around electrical danger are not electrical engineers who understand all the precautions. It's the laborer and the machine operator, and often the maintenance personnel who are getting electrical burns and shocks. There is an arc blast that can occur by just turning the electrical power on and off. Electrical injuries are the worst. I always have the smell of the burning in my head after investigating cases where someone stuck their hand inside or under machinery without locking out the electrical power. Sometimes they die if the electrical energy crosses their heart, but sometimes they live with the most devastating injuries. I have had some close calls.

I ask dozens of questions when I am inspecting machinery or going anywhere I have never gone before. I try to never lead the investigation. I let someone who knows where they are going be the leader, and I will cautiously follow. I used to get nervous when I would see OSHA compliance officers leading the way into places they had never been before. One time I saw an inspector reach for a light switch in a dark room that turned out to be a piece of machinery. I try to make the owners prove to me that the power is turned off and locked in the off position for any equipment to which I might be exposed. Several times in the steel mills, while climbing a fixed ladder, I have had my AC tester alarm and illuminate. My hand was inches away from certain death while climbing a metal ladder. I have done a considerable amount of teaching and training on electrical hazards. When I worked for the large electrical contractor, I trained five thousand union electricians. The trouble is after the training is done, I know that each of those workers can be tired, preoccupied, or make a single error, and they will be seriously injured.

MORE ELECTRICITY AND STATIC ELECTRICITY

Electricity is a natural phenomenon. It is quite mysterious. There were two great names of electricity. There was Thomas Edison and Nikola Tesla. Starting in the 1880s, these two were embroiled in a battle known as the war of the currents. Tesla believed that alternating current (AC) was the solution to managing voltages, and Edison was a proponent of direct current (DC). These giants of electrical engineering were both egocentrics, and they battled with each other for control over the electrical grid. These two were like the Einstein of electrical energy, and they seemed to understand electrical energy better than the rest of us. Neither of these greats was able to harness lightning or static electricity for use in our power grid. Both lightning and static electricity are powerful forces that represent danger in our world. Everyone has some fear of lightning and has seen its great destructive power, but static electricity is even more omnipresent in our world and extremely dangerous. You have probably had the experience of getting a static spark when touching a door handle, and you have also seen the magnetic power of static electricity when you have rubbed a balloon in your hair or on a sweater and had it stick to the wall. What you might not know is that static electricity sparks are of such short duration that they don't seem to do serious damage to your body, and the spark just hurts for a second, but that spark is hot enough to ignite flammable gas or vapor. This is a serious problem whenever flammable gasses or vapors are used.

The concept of grounding and bonding near flammable storage areas or when processes involve flammable material is based on the concept of controlling static electricity and sending it to the ground, where it can't cause a major explosion. When I see somebody using flammable materials or pumping gas, I might be alarmed by them smoking a cigarette, but the cigarette burns at lower temperatures than a static spark. Lighting a cigarette presents much higher temperatures than smoking a cigarette. It is unlikely that smoking a cigarette will exceed temperatures that will reach the lower flammable or explosive threshold for most materials. A cigarette burns at a few hundred degrees, but a match will be more than a thousand degrees. I am more alarmed by somebody pumping gas or using a flammable material while wearing a wool sweater. I am talking in generalities about the relative danger of an explosion, and I would not want the reader to think it is enough to just think of the potential for an explosion in general terms. In an actual situation, the user of flammable materials and gases must know the exact flammable range of the material and determine that there are zero sources of ignition in the area. The standards speak to at least twenty feet away from any source of ignition, which would always include static sparks and unapproved electrical equipment.

It has been my experience that the public does not have enough caution or concern about static electricity and does not know enough about grounding and bonding. If you are a hobbyist and painting with a flammable paint or other flammable material, you better know enough to assure containers are grounded and bonded to each other, or that little spark you feel when touching a door handle could send you to the emergency room or worse. If you are using any product that says it is flammable, you should use extreme caution. I have found hair spray to be one of the products labeled flammable. Too many explosions have been investigated where not enough effort was put into evaluating and protecting from static electricity. Another cause of those explosions was often the use of equipment that was not rated for a hazardous location.

There is equipment that is designed to be intrinsically safe and does not produce a spark that must be used in such environments. Forklift trucks are notorious for creating a spark that causes an explosion. There might be a factory that is fully grounded, bonded, and protected from static electricity, and somebody brings an unapproved forklift into the building, and there is an explosion. I have been to a number of facilities where flammable gas or vapors are used in a closed system, and there is a release to the general atmosphere and no detectors to shut down the process until it is too late to avoid a catastrophe.

TESTER AND DETECTORS

The equipment is essential to the safety professional. Whether it is to measure or weigh a load or the dimensions of a piece of evidence, testers and detectors are essential equipment for any safety professional. I always take a camera, including video, a tape measure, an engineering rod, an inclinometer, a light and sound meter, an electrical tester, and sometimes a heat-measuring instrument. I will also order specialized equipment when necessary and make use of testing laboratories when useful and necessary. I bring sample jars that I can label for evidence. I have acquired many kinds of scales. There is no time when instruments are more important than when I must determine if there is enough air for me to breathe or when I am involved with flammable vapors or gases. Those inexpensive and effective instruments will give warning to a situation where there is insufficient oxygen, or when there is excessive buildup or release of flammable gas or vapors. I have seen these instrument alarm on many occasions, and every time it happens, I think they may have saved my life. You might not expect to run into a situation where there is not enough air to breathe or where contaminants are at a level that could cause imminent danger, but I have to live in that world, and it has happened to me. It is unimaginable to me in these days of technology that there are thousands of car repair shops and painting operations without detectors. Spray booths can be used to contain dangerous vapors within the booth, but they are costly, and too often I have seen ignorant people trying to avoid spending money on such booths and just spraying flammable liquids, hoping for the best. When a combination of containment and detection is used, the likelihood of an explosion is significantly reduced.

RUN OVERS AND BACK OVERS

I believe if we had enough cameras, we could avoid most of these terrible incidents. There is a commercial that shows a car with cameras on the top of the car, and the passengers can see alligators below and around the car. The truth is the big trucks and construction vehicles have large blind spots, and everybody knows it and has known it for many years. The operators back up those trucks often without spotters or trained flag people, and it's a matter of good fortune that they don't run over people on a regular basis. One day, they can get a little too close to some unsuspecting pedestrian worker who loses a leg or a foot because of a lack of visual communication. This is a mistake that can change a life forever, and no operator should ever move their truck or equipment without being certain that the blind spot is clear. This is a problem with backhoes and excavators, where the operator's attention is focused on the dirt or the equipment they are moving, and since they don't have eyes in the back of their heads, two careers can be ended. The victim can't work again after being crushed, and the operator can't work after the shock and trauma of hitting another worker. It is also difficult for an operator to find work as an operator after such an incident damages their reputation.

After the incident, everybody talks about nothing but backup alarms or motion alarms. The question becomes one of whether it was working on the equipment and if it could be heard around the surrounding noise levels. The standards dictate a spotter or flag person when motion alarms cannot be heard over the surrounding noise levels. I think this is a completely wrong approach. First, nobody knows the levels that can or cannot be heard over surrounding noise, and second, I know from experience that after a few minutes working around this beeping equipment, it loses its value and effect. A person tunes out these alarms and goes on with their work under the assumption that nobody operating the equipment is going to run them over. I am not against having these alarms on the equipment. They are better than nothing at all, but putting them in the standards is a big mistake. With all the technology we have today, there is no reason that operators of trucks and heavy equipment should rely on these alarms. I want the camera and the flag people, and I also think that satellites could be used to identify the location of pedestrians. I have suggested that pedestrian workers wear bracelets that would give off a signal that would notify the operators of their proximity to operations. These bracelets and satellites would be especially effective during railroad operations. I have had too many cases where the railroad is pushing cars and locomotives, and they have no idea where their carmen, engineers, or other workers are located. If I was thinking of starting another safetyman business, I think progressive safety technology would be a good one. The same technology being used for production could be employed to prevent countless injuries and deaths. I think about the doorbell that is a camera that people have on the door of their homes. This doorbell connects to their cell phone and tells them when anyone approaches their home. The same could be used in the field on countless projects.

LADDERS

The city is having a painting event, and everybody is invited. A concrete wall separating the river from the railyard will be painted by artists. These artists will all paint graffiti on the fifteen-foot-high wall. The city and the group organizers are providing paint and ladders, and it is a free-for-all with artists climbing ladders fifteen feet in the air and painting their art. The only problem is that none of these artists have been trained in ladder safety, and none of these artists know how far the feet of the ladder should be away from the wall. To make things worse, the ladders are on gravel. There is a slope, debris, and holes. At one location, it was not possible for the ladder to be placed within eight feet of the wall. The young artistic climbs up the Fiberglas extension ladder and starts his painting when the ladder slides down the wall and he breaks his wrist. There are specific standards for training on the use of ladders, and it is difficult for a safety professional to understand how nobody cares enough to have any training or supervision of a group of young people climbing so high with no experience. It is interesting to me how unsafe most people are with potentially dangerous ladders. I know of many cases where somebody has been loaned a ladder that might be in bad shape or the person had no idea how to properly use the ladder. Ladders of all kinds are extremely dangerous and should only be used with an abundance of caution.

TUNNELING

I was the first and only tunnel coordinator at OSHA. It was after a tunnel explosion in Milwaukee that took the lives of three workers at approximately 9:00 a.m. on November 11, 1988. This was a Deep Tunnel Project for a water-pollution control project that started in 1981. The general contractor was cited by OSHA for inadequate ventilation. Events that led to the deaths that Thursday began about 8:30 a.m. when gas monitors detected methane in a twelve-foot-diameter sewer line thirty feet under the ground. Ten workers were evacuated, and fresh air was pumped into the tunnel to remove the methane gas that forms through the decay of organic matter. At 8:47 a.m., believing that most of the gas had been removed, workers reentered the sewer line, carrying gas detectors. They went down to see how much was still in the tunnel. Two explosions were heard at 9:03 a.m., and to this day, nobody has determined exactly what the source of the ignition was.

I was told the assistant secretary of labor was concerned about other tunnels blowing up and all the bad publicity from the explosion. Nobody at OSHA had much tunneling background, and I was the training coordinator. They asked me if I wanted to learn everything about tunneling, and I said yes, and they sent me to tunnel training in Canada, where they make boring machines. It was a great experience, and before it was over, I had the opportunity to visit almost every tunnel in the country. One thing that is important to understand about tunnels is that there is no air down there. If they don't have air forced down into the tunnel or you don't wear your air on your body, or, if for some reason something happens to the air, you will die. I remember the first tunnel I visited was with Shut 'Em Down. It was the deep tunnel under the city of Chicago, and it was three hundred feet deep, and it was both exciting and scary. I was surprised by how much action happens at three hundred feet. I was dropped down in a man bucket from a crane, and it was a very tough landing. Shut 'Em Down thought they gave us a tough landing on purpose because we were with OSHA. As soon as we landed, there were railroad tracks

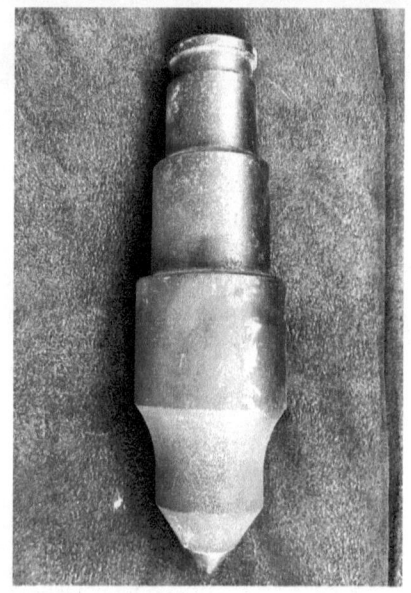

and a locomotive called a Loci running in two directions up and down the tunnel. I found out in a hurry that the locomotive was the boss of the operation, and it was my job to cling to the walls whenever they went by, so I didn't get run over. It was so cramped and so slippery that a few times I thought I might slip in front of the train, and that would be the end. Everywhere, there was water dripping and drilling in the roof for rock bolts, and the noise from the giant boring machine with the diamond cutters. I have one of those cutters in my living room that they gave me as a souvenir.

SIGNPOSTING

A worker fell from what is called a shepherd's hook; it hangs from the top of a billboard fifty feet in the air. You would think there would be

234

some secondary fall protection beyond a swinging hook at the top of a rusty old billboard, but this guy fell, not because the hook failed but because the actual top of the billboard broke away from the sign. He died, and when I investigated, I had to climb up on that billboard next to the highway on a rickety old ladder. I lost my balance and fell into the harness myself, but I was attached to a part of the sign that was more substantial than the top where the worker was attached. I can tell you it hurts to fall, and while I waited to be rescued, I was in a lot of pain. I am grateful that I didn't die and suffered no permanent injury.

I have investigated many cases where a worker has fallen in the body harness, and when not properly adjusted, the testicles are injured. I must say that if I had to do another investigation on an outdoor sign, I would use a lift. Many of you will not know that they used to use pieces of paper and glue to place signs on the billboards, and it was very labor-intensive. Today, most of the signs are made of a vinyl material that is placed on the billboard more like a fitted sheet on a bed. It is still dangerous, and they climb the ladder and use fall protection for the most part and do not have the benefit of a lift. I had a lot of questions about who inspects the structures of the billboards to make certain they are safe from an engineering perspective, and I did not get any good answers. They told me that they inspect the billboards when one of their workers reports there is some damage. Well, it is too late when they find out about structural damage from falling to the ground, and these sign posters are not structural engineers.

Another problem I found with signposting is the fact sign companies think posting of signs is a one-person job. That means even if you do fall, you might hang there and suffer or die. I recommended that my client provide a satellite tracking device and a signal that goes off immediately if there is a fall on any single-person signposting job. Apparently, nobody had thought about this until my investigation.

I sure hope things have improved since that investigation twenty years ago.

GRAIN ELEVATORS, SUGAR, AND OTHER COMBUSTIBLE DUSTS

When I started in safety, and especially at OSHA, there were many explosions occurring at places with fine dusts. When this fine grain dust, sugar dust, or any kind of dust explodes violently with just a single static electrical charge, it is like a bomb has leveled the factory. This problem was first identified in the coal mines. I remember as a little boy going to the Museum of Science and Industry in Chicago and down into the fake coal mine that I thought was real. They did a demonstration of a coal dust explosion, and it was loud and scary. Little did I know that I would be facing that same experiment in real life. Combustible dust explodes when fine particles are suspended in the air in an enclosed location. This dust can explode regardless of its size, shape, or chemical composition when suspended in the air at various concentrations.

When I worked for the state of Wisconsin, it wasn't long before I discovered there were grain elevators in northern Wisconsin that were deadly dangerous and exploding all too frequently. The last one reported was on June 2, 2017. It killed two workers, and the plant had previously been cited in January 2011 for not having an explosion protection system. There was one on August 22, 2014. I was asked to investigate the explosion in Savannah, Georgia, on February 7, 2008, where fourteen were killed and thirty-eight injured. There was one in East Rutherford, New Jersey, on October 9, 2012; Woburn, Massachusetts, on November 5, 2013; Vergennes, Vermont, on December 8, 2017; Winterville, Georgia, on September 23, 2015; Chattanooga, Tennessee, on March 11, 2015; Belcamp, Maryland, on

June 28, 2017; Flagstaff, Arizona, on September 14, 2014; Cambria, Wisconsin, on May 31, 2017; and many others. These explosions are still occurring at an alarming rate. If these were airplane crashes, there would be much more coverage and concern.

In my days at OSHA during the 1970s and 1980s, I was sent on a mission to Superior, Wisconsin, to stop grain elevator explosions. This is tough duty because they have not stopped to this day. What I found was the grain companies were not going to spend the millions for elaborate safety systems to prevent these explosions, and as far as I know, they still don't have those systems. Here is another example of the complex and difficult relationship between the cost and benefits of safety. In rural Wisconsin and other agricultural communities, spending millions for dust monitoring and filtering units just isn't in the cards, and the results are periodic dust explosions. I can remember my fear of inspecting these facilities because the only thing between me and oblivion was the housekeeping and broom sweeping of the plant, which is far from an exact science.

LAWN WORK AND CHEMICALS

Probably one of the most dangerous activities a homeowner can do besides cleaning the gutters is lawn work. Almost all the chemicals used are toxic and carcinogenic. You may have heard about one national brand in litigation that has been ordered by the courts to pay $80 million in damages because it was a "substantial factor" in causing non–Hodgkin's disease. This was a glyphosate-based herbicide, and there will be appeals in the courts. There are so many chemicals involved in herbicides with long, strange names like dichlorprop, mecoprop, dicamba, clopyralid, and triclopyr. These are just the weed killers. The insect killers have abamectin, cyfluthrin, fipronil, permethrin, bifenthrin, hydramethylnon, pyrethrum, and many others. We could create an entire book on each of these chemicals, and there would be lists of precautions and protections required for the user.

When you contemplate these complicated chemicals, you realize we need a specialist to deal with these potential problems. They will spend the time and money to learn about the danger and the necessary protections. Unfortunately, we don't want to spend money for these people and analysis and testing, so we go to the store, buy the stuff, ignore the labels, and spread it all over our lawn or all over the neighborhood. These chemicals cause cancer and have shortened plenty of lives, but there is no way for us to prove it until it becomes tragic news. I have observed some of the professional lawncare people failing to follow the rules and not wearing the proper personal protective equipment. As a safetyman, I would advise everyone using chemicals with these long names that you haven't heard, can't pronounce, and that come with two pages of warnings you didn't read to, at a minimum, wear chemical gloves and a good-quality respirator.

Humans should wear a canister respirator and full personal protective equipment. A paper dust mask is not enough. Once the work is done, make sure you have used disposable clothing and never wash that clothing with the family's clothing. The clothing worn during the application of these chemicals should be taken off immediately and stored in a secured, closed bag. I am aware of too many cases where workers were exposed to asbestos and came home from work with fibers on their clothing, exposing the entire family. I am a safetyman but not a perfect person. Despite my awareness, there have been times when I put fertilizer on my lawn and didn't read all the labels, but I could not sleep, thinking that my clothing might contaminate the home and family. Also, these chemicals are harmful to your pets and possibly your DNA. If I oversaw the world, I would not let homeowners deal with these unknown and potentially deadly chemicals. I would make us hire professionally trained and certified specialists. I believe it is wrong that we have more protection from hazardous chemicals at work than we do at home. We need a hazard communications program (see OSHA's Hazard Communication Standard) for every home. Chemicals would be fully analyzed, and the proper personal protective equipment would be certified.

At home, because nobody cares, it is the Wild West. The safetyman knows that people are not reading those labels and taking those precautions, and nobody is doing anything about it. This lack of caring and responsibility affects many other aspects of life. Gasoline is a hazardous chemical that is dangerous when in contact with the skin and can be absorbed through the skin. It seems reasonable that it would be prohibited to pump gas in your car without gloves. The gas pumps have vacuum systems that significantly reduce exposure to the gasoline vapors, but they do nothing to protect us from the skin-absorption problem. Every gas station should provide proper disposable gloves.

WARNINGS AND FEAR OF FALLING ICE

I am not against warnings, but they often don't do any good, and sometimes people think warnings are all that is necessary. Much like the case I mentioned about slipping on water in the hospital, every establishment must deal with this problem of a slipping hazard. Placing an orange cone in the general area of wetness is not going to solve this problem. Instead of placing the cone, somebody should be cleaning it up and looking at the situation and trying to remedy it permanently. I wish I had a dollar for every time I told somebody that they need to place a floor mat outside and inside a doorway to prevent an accident.

I get the feeling that I am the only person who cares about this, and unfortunately, I don't run the world, and many people think it's okay to let a few people fall. They certainly don't think that a fall will result in having a leg amputated or hitting their head and getting a subdural hematoma.

One of my favorites, or least favorites, is the falling ice signs I see in Chicago whenever there is a thaw during the winter. Ice forms on the skyscrapers, and when the weather turns warm, the ice falls to the ground and injures and sometimes kills people and damages property. On November 29, 2018, authorities had to block off the streets. On December 4, 2018, ice fell and smashed windshields, and large chunks fell from the Willis Tower. On February 12, 2018, the same thing, and this was just Chicago. When I go to the Loop and see the signs to "watch for falling ice," on the one hand, I think it is silly since nobody can watch for the falling ice and walk at the same time, and we can't walk in the streets because we would get run over. With no reasonable choices, I fear that my life might end at any moment. These building engineers have an obligation to anticipate this problem and devise a solution, and the city must do more than just tell the owners to put out signs that are better than nothing but don't solve the problem. There was news last night that a woman architect was killed yesterday from falling debris in New York City.

HUMAN HEARING

MANY OF US ARE LOSING OUR HEARING. THE LOSS OF HEARING
is a serious disability. It affects the quality of life, and it also affects
the safety of the individual. Sound cues are one of the most protective
means of identifying danger in our hostile world. I am not aware of
any statistics that have indicated how many injuries or deaths have had
hearing loss as a contributing factor or proximate cause. Our hearing
is sensitive, and we are only beginning to look at our children and
attempting to protect their hearing so it will last a lifetime. In the last
few years, small children at sporting events and concerts have been
wearing hearing protection. Most of the physiology of hearing has to
do with small delicate bones (ossicles) in the ear and hairs (cilia) that
transmit sound to the brain. These bones get more brittle over time,
and the hairs disappear. What we do know about the science of hearing preservation is that noise, and especially loud noise, will diminish
hearing ability at a faster rate than normal.

Concussions in sports and life affect our brains. We are getting
much more protective of our heads and our brains. I believe the same

will happen with hearing. When I go to factories, construction sites, and downtown into cities, there is a surprising amount of noise to endure. At a sporting event or a concert, I am the only person wearing hearing protection. Even on the airplane, there are only a few other people wearing noise-canceling devices. I believe there will be a major effort in the future to reduce or at least control noise. Noise waves are an attack on the human physiology and should be viewed like any other threat to safety and health.

There is a scandal going on at the present time about earplugs supplied to our armed forces that were not as sufficiently protective as advertised by the manufacturer. The quality of hearing protection is an important discussion because when it comes to protecting hearing, one size does not fit everybody. There should be a scientific approach to protecting our hearing that goes beyond handing out earplugs. The standards for hearing protection should be reevaluated. Currently, standards recommend engineering controls to reduce noise, but they have been accepting earplugs as a compromise. The standards for the protection of human hearing should be just as strict as those protecting the body from the point of operation of a machine. The equipment, machinery, and environment should never be allowed to damage our hearing, and personal protective equipment should only be used after engineering controls have been implemented.

In the home, the vacuum cleaner, lawn mowers, and leaf blowers are dangerous when it comes to noise. I see people using these and other noisy equipment without hearing protection. I have a sound level meter on my cell phone, and I have calculated levels as high as 110 dba. That reading is comparable to standing next to a jet engine or firing a handgun. This is not acceptable, and the manufacturers of these products must be required to reduce the noise or provide proven protection with their equipment. The problem with the manufacturers of any of this equipment solving the problem by providing hearing protection is that there is a hazard created when the person operating the equipment can't hear what is going on in the environment. I have had cases where someone riding on a sit-down lawn mower backed into another person who was screaming at them to stop. We need to develop technology that reduces the noise and still allows the operator to communicate. Some of this technology already exists and can be seen on the football field, where the quarterback gets communication from the sidelines, but the crowd noise is screened out.

SCHOOL BUSES AND LOOKING AFTER OUR CHILDREN

We are not doing enough to protect our own kids. I don't have to tell parents that kids are just not as reactive, sophisticated, or knowledge-able as adults in emergency situations. In a fire or evacuation, adults will find the exits and have had some experience in the proper actions necessary for self-preservation. Young children depend on adults for their survival, and we don't always take their lack of experience and development into consideration when planning for potential danger. You would think this would be a priority. I can't believe how bad some school buses are constructed and maintained. They have no seat belts or airbags, and the way they are built with the rear end hanging far beyond the rear axle, no other vehicle would be allowed to exist in that configuration. I have investigated some school bus injuries, and in one case, the alleged emergency exits amounted to a doggie door in the back of the bus that didn't open easily in an emergency. There were also supposed to be cameras that were directed at the driver and the children. The camera wasn't working. There have been cases where there was not good traffic control between the children and the school buses or between the children and the parking lot and cars. Children do not always stop, look, and listen before they run toward another child or an adult. There should be detailed plans for safeguards at schools and parking lots that include provisions for a traffic control plan (TCP). These include signs, signals, traffic guards, and speed bumps. A traffic engineer trained in safety should be involved in designing and planning traffic control in areas where we know children will be present. When I see twenty-mile-per-hour signs in front of a school zone and no traffic control plan directly in front of the school drop-off area, it makes me fear for the lives of those children. I have noticed that there is sometimes better traffic control at airports and shopping centers than at elementary schools. Where children are involved, we should be putting forth the time and expense to investigate their safety and health.

ELDERLY AND INFIRMED

Every precaution suggested for children should also be taken for the elderly. Many older people are even more limited in their ability to react to an emergency or a hazardous situation than the children. The children in school may not need the kind of physical assistance needed by the elderly and infirmed. Planning is needed not only to move the people but also the equipment, such as oxygen tanks, walkers, and hospital beds in and out of the facility. As our population is aging, more places are being developed for seniors, and those facilities need evaluation by safetymen and safety women.

FITNESS

The fitness and the wellness movement have been great for improving health and extending life, but there is too little information about when enough is enough. There is also a big void in training people on how to use the equipment. This safetyman has seen many injuries from people not being trained on how to use the treadmill, the elliptical, or even the weight racks. It is not enough to sign people up for the health club and show them the equipment and let them loose to their own devices. There must be an orientation and training program to teach people the limitations of the equipment and the limitations of their bodies. There should be training programs and coaches for running, especially for marathons. Thousands of people are running marathon races without their doctor's care or proper training.

BOATING AND MARITIME

Boating accidents are common today on crowded lakes and rivers with overpowered speedboats and little traffic control. Whenever I hear of a friend going boating or white-water rafting, I am always concerned. I have lost some good friends in those lakes and rivers. I had one friend fall between the boat and the pier, and another hit his head on a rock when falling out of a raft. OSHA regulates maritime operators but not recreational activities. I have had learning experiences during maritime inspections. One was on a casino boat where the shipmates were trying to move a large guard railing from the side of the boat, and one of them threw out his back. This is a common problem in all industries. They didn't weigh the railing and had six people lifting this irregular, odd-sized piece and ended up ruining this one guy's back. He had been trying to hang on to his end with five other people, high winds, and rocking waters.

Another time, I inspected a concrete hauling vessel on Lake Michigan. This was an eye-opening experience about the dangers involved in shipping over the high seas. A worker had been caught and killed in a huge concrete conveyor that went from end to end on this enormous ship. I found there were zero safety programs on this ship and was told it was custom and practice to not practice safety in shipping operations. This large, completely unguarded conveyor system the entire length of the ship had no lockout program to make certain that when the conveyor was started up, nobody was inside the conveyor. Unfortunately, this seaman was making an adjustment, and they turned on the machine. I had seen this type of death previously at a railroad wheel-manufacturing operation when a man was crushed in a wheel conveyor. I had also investigated a man being crushed at a car-crushing operation where they had a conveyor system underground. In each case, there was no lockout procedure and no emergency stop buttons or wires within reach of the fatally injured workers. I learned in all these cases that nobody was thinking about safety from the standpoint of understanding that one mistake with a conveyor system was certain death. Also, I learned that these lengthy machines, such as conveyors and printing presses, have the problem of people not being seen when they start up the equipment. Accountability of all the people before starting up the operation is a big part of the lockout standard.

The concrete dust on this conveyor ship was thick within the entire hull while the concrete conveyor was running. Breathing silica dust was not what I expected when they demonstrated the conveyor system to me on this ship. The entire area was one big dust storm, and I could hardly breathe. I have no doubt that the single exposure to that concrete dust shortened my life. I asked about any industrial hygiene air sampling or a respirator program for the ship, and they looked at me, shaking their heads. There was also the unforgettable noise. The noise was unbelievable from that conveyor when it was operating. I didn't get a chance to measure the noise, but it was like being directly in front of a jet engine. I would estimate it was over 180 dba, and that would be twice the allowable limit. I was glad I wasn't doing maritime inspections on a regular basis. Many of my retired OSHA fellow compliance officers have respiratory problems and are wearing hearing aids.

WALKING INTO GLASS

Almost everybody I talk to has done it. If you don't put a decal on glass, many people will not see it and will walk right into it. I saw a video on YouTube about people walking into glass. It was supposed to be funny, and the idea is people who walk into the glass are stupid. The guy on the video sees a woman walk into the glass door and says, "Well, she won't be able to reproduce more stupid people." I can't believe this video exists. Watch it, and you will see the problem with safety in our world. There is no mention of the fact that it is hard to see the glass. There is a glare, and many people don't see well. It's a cold world. People have had life-changing injuries walking into glass or walking through glass if it is not thick enough or not safety glass. I had a case once where a young man got locked out of his apartment and tried to get in through a basement window. The glass on the window was so thin that when he touched the window, the glass broke, and he got a very severe laceration, losing the use of his dominant hand.

There are many standards and codes for glass because it can be a significant hazard. I remember working with returned bottles when I was a teenager at the grocery store. Every week, somebody would get lacerated by the broken glass. One day, they handed out leather gloves for those of us dealing with the returned bottles. The manager was complaining about how much the leather gloves cost, but the lacerations ended. Safety glass was an important advancement, not only for panels of glass but also for safety glasses to protect our eyes. Tempered glass processed by thermal or chemical treatments increases the strength, and it will crumble into small granular chunks when broken instead of jagged chards. There are several ways for a safetyman to identify safety glass. Usually, it has a marking, but the best way to tell is to look at the glass through polarized lenses. The polarized lenses will show dark, shady spots or lines stretching across the surface. Anytime a window or any glass surface needs to be repaired, tempered safety glass should be used.

FALLING FROM THE STAGE

I will never embarrass my client, who was a leading attorney, specializing in dealing with OSHA. She was nice enough to invite me to make a presentation in front of at least two hundred other attorneys about OSHA. It was held at a major hotel in Chicago, and it was a big opportunity for me. I was a little nervous standing behind the podium, and she started the presentation by giving me a big buildup and introduction. She had come from the side of the stage, which was at least five feet above the floor. She gave me the introduction, and as I was looking at my notes and getting ready to say my first words, I heard a bang and a giant reaction from the entire room. This lawyer walked straight off the stage, completely missed the stairway, fell to the ground, and sustained a serious enough injury that the paramedics had to be called. After she was given attention, it was my job to give a safety and OSHA talk. It was a difficult situation, but it was also the perfect time to talk about the importance of safety and the human condition. We talked about the stage not being guarded and that there was no information at the location of the stairway. Most important was the human factor of someone being completely distracted.

I must admit I have seen several other people fall from the stage. I don't think it's that unusual. This idea of walking right off the edge is too common, and it is mostly because, on a stage, the "actors" are nervous and distracted, which is a dangerous combination. I have seen a distracted person walk down a ladder backward, facing away from the ladder even though that person had used a ladder properly, facing the ladder, many times in the past. I have seen several times in construction where two workers pick up a piece of plywood, and one will walk right into the hole the plywood was covering. This is the reason the plywood cover is required to be secured and labeled.

DENTAL OFFICES, MANICURES, HAIRSTYLING, AND HOSPITALS

These are scary places for the safetyman. I have been teaching blood-borne pathogens for thirty-five years, and I know all these places are required by law to have a written program for hazard communication, blood-borne pathogens, and specific procedures for the protection of their employees and the public. The first time I went to a dental office and asked for a blood-borne pathogens program, they looked at me like I was from Mars. Since it was my longtime dentist, I got them to hire me to conduct a survey and write a program for them. I went to another dentist and an endodontist, and neither had any program or procedure to protect me and others from hepatitis, AIDS, MRSA, and many other potential blood-borne health hazards. We never even talked about airborne hazards like tuberculosis and Ebola.

The same is true for hospitals I have visited. The hospital may have some programs, but if you ask for them, they get angry, and the alleged professional people you ask about these issues seem to know very little or nothing at all. Try asking your own health professional about MRSA. I used to like manicures before I learned about the danger of blood, bodily fluids, and lack of sterilization and training. The last time I went many years ago, I realized that there is almost always blood involved, and I am dealing with a person who can't even pronounce blood-borne pathogens. I will do my own manicures.

The same when I get a haircut. I go to several nice places, and when I go to these places, I like to ask a lot of questions. When I ask questions about head sores, head lice, and sterilization procedures, I don't always get straight answers. Sometimes I ask a stylist if they have seen any sores, head lice, or mistakes with the sterilization procedure, and when I look at their faces, I think they are telling me yes, they do see that stuff. It makes me watch them more closely.

Recently, a friend told me his wife got herpes all over her body in the hospital, where she was having thyroid surgery. He told me they determined the thyroid surgery was unnecessary, and now she has herpes. I asked him if the hospital admitted they had performed unnecessary surgery, and he smiled and said, "You know they would never admit to that."

There are too many mistakes being made by individuals who are supposed to be looking after our safety and health, because people don't appreciate the danger. I can remember taking my daughters to the mall to have their ears pierced and realizing they had no sanitation procedure. Usually, there are procedures to protect customers and patients but no consistent effort and no real understanding of the dangers of blood-borne pathogens and other essential issues dealing with disease and danger in their workplace. The only place I even see elaborate procedures and protections is at banks or jewelry stores where they are trying to protect their assets, and even those places sometimes get robbed.

Then there are the needlesticks. I found out that many dental workers get stuck by needles when anesthetizing the patients. I started asking them if they ever get sticks, and they all said yes. At the dentist where I did my survey, they said they had seven needlesticks in the last three years. These needlesticks require years of testing to assure the worker has not contracted HIV, yet none of that was happening. It is scary living in a world where you know about these blood-borne and airborne dangers.

Then there is measles, which is now in the news, and many other health hazards discovered every day. One time, I was in a restaurant bathroom, and I reached up for the knob on the paper towel dispensers, and there was a sharp burr, almost like a knife blade. It struck me, and I bled for five minutes. I looked at the burr or blade, and I could see dried blood where other customers had been stabbed at the same location, and I realized I was playing Russian roulette with my health. I might die, or I might be fine. Who knows, and nobody cares. When I tell people about this incident, they smile and tell me I will be just fine, like they are sorcerer scientist and I am a child. Maybe it's just they don't want to be bothered thinking about my problems for even a second. All these people get sick or die, and they never even learn how they contracted their diseases. This is the great nightmare of the safetyman.

DISTRACTION AND HABITUATION

With all our technology, distraction is a big problem. When we are distracted, we are not in a good position to protect our safety. The problem with habituation is we have found that once a person gets in the habit of taking a certain path or doing a task in a certain way, they will continue to do the same thing repeatedly. It is like the students in my class who always take the same seat, without thinking of the reason. Even if we put an obstacle around those seats, those students won't even notice the difference until they are determined to return to their same seats.

In my graduate work at a meatpacking plant in Madison, Wisconsin, we watched and videotaped how people would walk from their parked cars to where they would went into the plant. Each day, we would place obstacles like water or stones to see what it would take for them to change their course. If we put rocks in the way, they still took the same course. If they had to jump over a water puddle, they still took the same course. Even if we placed a strange machine that they needed to walk around, they maintained the same course.

It is so important to take habituation into consideration during construction activities. We must take precautions for safety even when obstacles are introduced into the simple task of walking to work from the parking lot. There is a human condition called negative transfer. It is a common experience. I have one car with standard transmission and one with automatic transmission. When I get into the automatic transmission car, I look for the clutch, and when I go from the automatic transmission to the standard transmission, I often forget to set the parking brake, which means my car can roll. I am a careful person, but I have seen my car roll after I exit and have had to run into that moving car to set the brake. One time, I even got a little dent in the door when I parked the standard transmission car at the gas pump, forgot to set the parking brake, and it rolled into a metal rail protecting the gas pump. I didn't feel like a safetyman when that happened, but I am a human just like everybody else, and negative transfer is a real problem.

TRAFFIC SAFETY

Traffic safety is where we all learn the importance of safety rules and regulations. People are taking chances with their lives, and we have incompetent drivers and bad traffic control. There are some traffic configurations and control plans (TCPs) that send you into trouble when they are confusing. The enforcement program of traffic laws is lax. The only time we see traffic control in my area is when they set up a roadblock to check for seat belts. They collect a few seat belt violators, a few drunk drivers, and a bunch of people driving without a license.

I believe the trucks should be separated from the rest of the traffic, like trains. When I was younger, the trucks were restricted to the right lane, but now they are speeding past me in every lane, and many of them have tandem loads swinging around outside the lane. The situation with trucking needs to be improved.

Another problem is when construction activities or construction crews impede traffic control. This is far too common, and there are too many situations where the construction equipment and vehicles fly into the lanes of traffic, where an alert driver might swerve out of the way, or an inattentive driver will run into the equipment. Trucks with oversize loads strike the viaducts and temporary power lines that are sometimes too low. There are traffic control signs that are so confusing that cars and trucks drive right into the barriers. There is room for traffic safety improvement in this country and infrastructure. I fear driving over a dilapidated bridge, and the potholes are destructive. Traffic safety is important to all of us, and we must do a better job. There are some good laws for traffic safety in the Manual of Traffic Control Devices (MUTCD). These rules come from the Department of Transportation and have been adopted by OSHA and many other places. The people who wrote this excellent standard did a great job, and now we must follow it.

DEMOLITION

It would be easy to have engineers conduct a survey of a building or structure to determine what might happen if things are moved or changed. If you are taking out a door or a wall, something surprising might happen. Also, if you are going to blow up the building or take down a beam or roof from overhead, it only makes sense to make certain the people are not in the structure or under the load. It's also the law. The OSHA standard is clear about requirements for a demolition survey, but rarely does a safetyman find a demolition survey has been conducted. They either don't understand the concept or don't want to pay for a qualified engineer. There is a case where they were removing steel beams for reconstruction. The beams fell onto people who were doing the work. Major construction and demolition firms are conducting demolition activities without proper planning and consulting with an engineer.

FIRE AND FIRE EXTINGUISHING

Once, I had a situation in my fireplace where the fire was starting to burn the fireplace mantel. I took a fire extinguisher, pulled the pin, and then when I shot it on the fire, the fire blew out of the fireplace and onto my oriental rug in front of the fireplace. I was able to easily put it out, but I damaged an expensive rug. Another time, I had a grease fire break out in an oven at an old apartment. I had a fire extinguisher, and when I pulled the pin and aimed it over the glass on the front of the oven, suddenly the glass on the front of the oven exploded when the cold fire-extinguishing material hit the hot glass. The lesson here is that I need more experience with fire extinguishers. The first time, I was too close to the fire, and the force of the extinguisher pushed the wood out onto the rug, and the second time, I should have opened the oven door, but it was too hot, and I feared getting too close. One other time, dry weeds had grown around our backyard firepit. It got too hot and was spreading over the firepit and onto the lawn. We had to isolate the fire to prevent it from spreading. My admiration for the firefighters who must deal with these kinds of problems every day has grown after these experiences. Everyone should have a properly sized and rated fire extinguisher in their home, but it is just as important to have some training and experience. It's not a good idea to have to use your fire extinguisher for the first time in an emergency. There are plenty of classes available at a local community colleges or at the local fire department where you live. Gaining some knowledge and experience in firefighting could save your life. Below is the firepit we use that is dug out, piled with stones, and with no weeds.

WIND

The wind is a natural phenomenon. It has a huge effect on safety and health and not too many established rules. There have been catastrophic incidents caused by the effects of wind. Everybody has had their umbrella pulled right out of their hands or turned inside out and has experienced the power of nature. The problem is that this power and danger is uncalculated and unpredictable. There have been notable catastrophes. There was a complete stage failure at a concert in Indiana where many were killed, and it also happened in Brazil in 2017. Building materials fly around at high speeds and destroy a construction site and injure people. A forty-mile-per-hour wind will fell whole trees. Severe windstorms are difficult to predict, with some gusts come with no warning. Wind advisories start at thirty miles per hour. Rules generally state that work should stop, and people should take shelter, but this is rarely enforced and a difficult problem at public outdoor events. There is a great need for better standards and enforcement in the area of protecting people from the wind.

RAILROAD ACCIDENTS

Railroads are far behind the times with safety and health. The trains are old, and the signal systems have not been upgraded in decades. The railroad crossings are just waiting for some unsuspecting driver or pedestrian to make a wrong move, and there are frequent accidents and injuries. Not far from where I live, there was a bus-train collision at a typical crossing where there is a small hill where the train crosses the road. In this case, seven students riding aboard a school bus were killed, and twenty-one were injured. The school bus was being operated by a substitute driver. The substitute driver had stopped at the traffic light, with the rear of the bus extending onto a portion of the railroad tracks. This rear portion was struck by a train on the way to Chicago. The tracks and signals for the rail crossing were located close to the highway intersection and traffic controls. The traffic lights and crossing warnings were supposed to be timed and coordinated to prevent such collisions. The investigation revealed there had been complaints from the public about the insufficient timing of the warnings provided by the signals in the year prior to the crash. There were some reports of near misses that had occurred prior to this incident.

The National Transportation Safety Board (NTSB) investigation found that the bus driver was not aware that any portion of the bus was on the tracks. The NTSB also found the timing of the signals was so bad that even if she knew the back of the bus was on the tracks as the train approached, she would have had to go through a red light to avoid the crash. After the incident, there was some legislation and reengineering to prevent a reoccurrence at this intersection and other intersections around the state of Illinois. There were also some recommendations for improved bus driver training. This train crash has been described as the worst crash in history.

SLAUGHTERHOUSE

I spent a lot of time in the meatpacking industry and became acquainted with many serious issues relating to both safety and health. There are problems with carpal tunnel syndrome, machinery injuries, and slips and falls, and there are some health issues. The problem with carpal tunnel is serious. It exists in the long line of butchers, conducting boning of animals with their hands and wrists. Sometimes the bones are removed with high-pressure machines. One time, I went to a chicken-processing plant in California, and they had a machine under pressures that shot out a part of a chicken, and it hit a worker's head, and he had a significant brain injury. At a slaughterhouse near Chicago, I spent more than a month trying to improve their safety effort. Dealing with live animals is a real challenge, and sanitation and processing are hazardous jobs.

I had a case in the meat-processing industry at a kosher meat-processing facility. At this plant, the rabbis inspect the meat and stamp the meat when it is approved to be sold as a kosher product. The rabbis climbed ladders to get up on elevated stands to stamp the meat. The rabbi was stamping tongues as kosher, and he fell from his stamping ladder stand and landed on his head. He continues to suffer a debilitating brain injury. The ladder stand was outside the normal standards customs and practices, and there was insufficient fall protection from the platform. In addition, the entire process required the rabbi to reach out on a constantly moving assembly line of beef tongues in an awkward position.

One time, while working for a Midwest state government, I became involved with the issue of listeria. There were questions regarding the lack of some significant procedures for decontamination that were absent in the industry. There was a sanitation issue regarding what is known in safety as "deadman controls." There is an automatic shutoff when the handle of high-pressure water is released. For safety reasons, it was determined that these hoses with the high-pressure water must have deadman controls because if the hose does not turn off when released, it can fly around and injure workers. On the other hand, the health inspectors felt that there was danger of difficulty in cleaning the mechanism that operates the deadman control. My efforts were to convince the health people that the protection provided by the deadman control was worth any additional effort needed for cleaning the equipment. Dealing with the meatpacking industry, a safetyman realizes how important safety and health programs are to our food supply. Overall, it is amazing how sophisticated the safety and health programs are in this industry. It is a complex place, even more complex than the steel mills and refineries that I have visited. So many different potential hazards in one place, starting with dealing with a live animal and ending with near perfection in cleanliness.

VIOLENCE IN THE WORKPLACE

I HAVE HAD TWO CASES. ONE WAS IN ALASKA AT THE PIPELINE. A worker was so angry and dangerous that everybody was afraid of him. He would beat up his fellow workers for no reason, and the bosses at the pipeline were afraid to fire him because he would come after them. Finally, when he almost killed a worker, the police got involved, and he was put in prison for a short time. My job in that case was to represent a worker who was beaten up by this guy and his claim that his employer, the pipeline company, had a duty under OSHA to protect him from a danger they knew and understood existed at their workplace. OSHA has a workplace violence standard that employers must act when they know there is the possibility of workplace violence, and if they do nothing, they have to pay an OSHA fine.

Another case I had was with the railroad. A worker was angry about being suspended and expected to be fired. In the middle of the night, he got his car into the rail yard, without any screening, and he started ramming into cars in the parking lot and hit one of the workers who was in the lot and some others who were in their cars. The police intervened, and the guy was put in jail for a short time. It was my job to tell the court that there is an OSHA standard, and the company should have known this guy, whom they were suspending and firing, might be angry about it. Here is what I wrote about the railroad case, with the names changed and the OSHA standard reference:

On or around June 16, 2005, at around 9:00 p.m., Mr. John Smith, a railroad worker, was seriously injured in Wisconsin yard. Mr. Bill Crazyman was able to enter the rail yard without being screened or identified and operate a vehicle without the consent of the owner at a high rate of speed while intoxicated within the rail yard. Mr. Smith was conducting switching operations and following his yardmaster's instructions near the roundhouse runner switch when the intoxicated (according to the police report) Crazyman entered the property, went into the employee parking lot, and allegedly was stealing a Pioneer stereo. He was able, without being noticed, to climb into a 1996 Jeep Cherokee, green in color, which belonged to an agent of the railroad. Mr. Crazyman stepped off the engine upon which he was working, called the tower asking for the police, and was waving his arms in the air to motion for the vehicle to stop when the vehicle came right at him. He ran to get out of the way, and the vehicle swerved in an apparent attempt to hit him. Crazyman jumped out of the way of the vehicle and struck his knee on a pile of stacked rail. He and Mr.

Worker (another rail worker) were then able to stop the vehicle and overpower the operator. There were no fences to restrict access to the employee parking area, no security cameras, and the telephone system directly connected to the police department was not operating. There are hazardous materials, including explosives, ammonia, and other toxic materials in the yard on a regular basis.

The following is an excerpt from the OSHA webpage about violence in the workplace:

Standard Interpretations
12/10/1992—OSHA policy regarding violent employee behavior.

Although currently there are no specific Federal OSHA standards to address these problems, the Federal Occupational Safety and Health Act (OSH Act), in Section 5(a)(1), provides that "each employer shall furnish to each of his employees employment and a place of employment which are free from recognized hazards that are causing or are likely to cause death or serious physical harm to his employees." In a workplace where the risk of violence and serious personal injury are significant enough to be "recognized hazards," the general duty clause would require the employer to take feasible steps to minimize those risks. Failure of an employer to implement feasible means of abatement of these hazards could result in the finding of an OSH Act violation.

The US Bureau of Labor Statistics shows that homicide is the second leading cause of death on the job. According to the Department of Justice National Crime Victimization Survey, approximately 2 million persons are threatened or assaulted each year at work. This includes violence by co-workers.

WORKPLACE VIOLENCE DEFINITION

According to the National Institute for Occupational Safety and Health (NIOSH), workplace violence is any physical assault, threatening behavior, or verbal abuse occurring in the work setting. A workplace may be any location, either permanent or temporary, where an employee performs any work-related duty. This includes, but is not limited to the following:

- the buildings and the surrounding perimeters, including parking lots;
- field locations;
- clients' homes; and
- traveling to and from work assignments.

Many types of behavior can disrupt the workplace and result in violent outbursts, including the following:

- verbal threats to inflict bodily harm, including vague or covert threats;

- attempting to cause physical harm—striking, pushing, and other aggressive physical acts against another person;
- verbal harassment—abusive or offensive language, gestures, or other discourteous conduct toward supervisors, fellow employees, or the public;
- disorderly conduct, such as shouting, throwing or pushing objects, punching walls, slamming doors;
- making false, malicious, or unfounded statements against coworkers or supervisors;
- inappropriate remarks, such as making delusional statements;
- fascination with guns or other weapons, demonstrated by discussions about weapons in an inappropriate context, or bringing weapons to the workplace.

ZERO TOLERANCE FOR VIOLENCE

One recommendation for minimizing the threat of violent acts in the workplace is to develop a "zero tolerance for violence" policy. This policy should convey that any type of violence will not be tolerated by any employees, customers, suppliers, or visitors to your establishment. This type of policy may be incorporated into the company safety and health program or an employee handbook. Possible content for the policy might be the following:

- a mechanism for reporting violent incidents;
- a method for disciplining violent acts;
- proactive anger-management or conflict-resolution training for employees;
- a resource person or team for handling violence complaints;

- an employee assistance program.

Because employee morale, productivity, and well-being are essential to a healthy workplace, it is recommended that employers provide an employee-assistance program (EAP) for employees who wish to discuss violence and other concerns in their lives. An effective EAP promotes healthy and productive employees, which is beneficial to all levels of personnel. Morale may be improved, and communication is strengthened between management and employees (source: Minnesota OSHA: http://www.doli.state.mn.us/wvprisk.html).

Workplace violence is a serious problem, and the numbers have accelerated in recent years. The latest was on February 16, 2019, where a gunman worker killed five people at a manufacturing business. The shooter used a Smith and Wesson, and the incident lasted more than ninety minutes until the police could take the shooter down. He was scheduled to be fired that day. Some of his fellow employees described the shooter as "friendly." When workers go crazy without any warning, this becomes an especially difficult safety and health problem. The company can implement all aspects of the OSHA policy above, and there continues to be violence in the workplace.

SPORTS SAFETY

I serve on the ASTM F10 Committee for sports equipment safety. There have been a lot of improvements over the years. Thirty years ago, there was very little personal protective equipment used to protect players, and there were a lot more injuries. There have been terrible injuries, especially in hockey with the flying puck, moving faster than the eye can see and doing significant damage to the players and the observers. There have been some improvements with the netting extended to protect the fans. They have extended the netting in baseball but not far enough to protect most fans. There is a move currently to expand the netting to the outfield. I used to think it was nuts to hear an announcement that the fans at hockey and baseball games should keep their eyes on the field or the puck and ball for their protection. As a human factors expert and a fan, I knew it would never work for two reasons. One is that the puck and the ball fly so fast the eye can't keep track of the projectile, and second, humans cannot possibly keep their focus on the game or the object continuously.

There are few studies or statistics about these problems in sports. Statistics must be gathered and an analysis of hazardous locations at the sports venues must be conducted. Even the players get hit and hurt. The answer to safety and health is to devise engineering controls, and if that isn't possible, then we must protect the people with barriers and protective equipment. Some people complain that the protective nets interfere with the vison of the game, but contrast that issue to a serious injury that can change your quality of your life, and it seems like a small price to pay. The worst is when I see parents bring their young children to the games, especially a hockey game, and not sit behind the netting. When I see this, it reminds me of Michael Jackson dangling his son over the railing of a hotel. It might be all right, or it might turn out badly.

I know from personal experience that by the seventh inning or the third period, many fans are fatigued or impaired and in no position to deal with danger. A few fans fall over the guardrail, trying to catch a baseball. I won't risk my life for a fifteen-dollar baseball. Those guardrails at the stadium, especially on the upper deck, are not up to code. A guardrail must be at least forty-two inches, and when I go to stadiums, they are never taller than thirty-six inches. I suspect it's to protect the view of the seated patron. This is a dangerous violation when those fans are standing. Somebody should investigate this. Too many sports refuse to include safety in the standard operating procedures. Look at soccer with all those headers and ice-skating and gymnastics all with the potential for serious brain injuries. Tennis is just starting to get into eye protection. The area of sports safety is fruitful for the safety professional because it is an example of how our society discourages and even fights against progress in safety. It is more important for the ice-skaters to be beautiful and elegant than to be protected from injury.

MORE SAFETY DRAMA

Airplanes are crashing because the software or hardware was not properly tested, and the pilots were not trained on a new system. We believe that other people are looking after the details and doing the extra work to protect our safety, and then we find out that they forgot to do it. Safety still is not the highest priority. There is hypocrisy when they tell us that safety is their highest priority, as those are just words and are not put into action. Yesterday, a cruise ship in Venice, Italy, slammed into a post. People are saying, how could this happen? What they really mean is that it never happened before. I don't know the details, but I have been to Venice, and it is a city that is floating on water, and water, especially in the ocean, is unpredictable. One little mistake or lack of judgment could certainly cause the ship to crash into the land. I am not a sailor, but I am a fisherman. I have been on many boats, mostly small fishing boats, and I have seen the difficulty even the most skilled seamen have in landing that boat at the dock. I know that I am a safetyman and therefore the most serious person about safety, but we all need to be more serious about it because there is so much at stake. When I refer to the drama of safety, I really mean drama, because there is a constant interaction between death and destruction and the call for action. There is no greater area where the drama unfolds than in gun safety.

GUN SAFETY

No other issue brings so much emotion to the conversation than the one about gun control, which I prefer to call gun safety. As a disclaimer, I don't have anything against guns. I know they are necessary, and in the right hands, they are important protection. I want to be protected by guns, and I want those guns to be machine guns if somebody is trying to kill me, my family, and my friends. It is just that they are dangerous, and like any other dangerous material, there must be a policy, program, plan, training, and enforcement to make them as safe as possible. I do not understand why gun lovers and gun haters would not all be interested in gun safety. It would be good for everyone.

I could write a gun standard in five minutes. The standard would use engineering controls to make certain guns were safe. Guns would be locked up just like a dangerous chemical. Owners would need proof of ownership before they could be used, like an optical scanner, or today they have face scanners. Nobody can have a gun unless they go through proper screening and training. Imagine (John Lennon) if, in the beginning, we had required gun owners to go through screening and training. We would be in a much better place. Everyone can have a gun or two, but they need to have background checks, and they need to lock them up.

THE SYMPHONY OF CONSTRUCTION AND GENERAL INDUSTRY

I HAVE REFERRED TO THE WORK ON A MULTIEMPLOYER WORK site as a symphony. The work and the safety program must be coordinated from the top, and this coordination is just as important for production and the safety and health effort. Good production efforts are the same as good safety and health efforts. To have the project or the process work at a highly productive level, planning is necessary, and the tasks must be organized and evaluated. The same qualities needed in the safety and health effort are needed for any successful business. If the safety and health effort was truly embedded in the production elements and the safety and health gravitas was the same as the production gravitas, there would be fewer accidents and injuries. Coordination and planning is a cost-saving and productive methodology.

A productive area can be proper staging of the jobsite or facility. When planners and engineers design a plant or a jobsite, they do it in the most efficient manner. They don't want to lay down or stage material in an area where it will be difficult to get them out when they are needed. They don't want to build around laydown or staging areas in a way that makes it difficult or impossible for the crane to reach and utilize those materials. This is like being painted into a corner. It's the same for the safety and health program. The planners and organizers must make certain the work is being conducted so that one contractor is not working under the work of another contractor. They need to make sure that the contractors with the heavy equipment don't store or operate the equipment under the energized power lines. They need to establish access roads and aisles to accomplish the work. Too many injuries occur because the conductor of the symphony of construction did not establish proper access roads or aisles.

Many injuries occur when construction workers park their trucks on rough ground and trip and fall on the way to their work area. The conductor of the symphony must assure roads are clear and smooth and the porta-potties are in the right locations. I saw porta-potties hit by a crane when they were being utilized. There must be enough fire extinguishers, and everybody must understand the game plan. Questions or concerns must be answered before moving forward. Many accidents and injuries could be avoided if the contractors would check in with the general contractor before they take some action. Too often, they start their work without alerting or advising the controlling contractor. When a building or home is constructed, there are going to be thousands of holes that will be cut and covered to prevent falls. This cutting and covering of holes requires constant coordination and communication. Just one unprotected hole could cost a worker his life, and his family will be affected forever. Companies can never be casual about holes and guardrails in the workplace. They can be the most important and cost-saving work on the entire project. Accountability for cutting, covering holes, and floor openings is essential. Sometimes one contractor will leave a hole or floor opening open, and another contractor falls to his death or has a life-changing injury. There should be a permit system for holes and floor openings. No hole should be cut without a permit, and the permit should be checked for compliance and safety. There is no better place for a permitting system than for floor holes and openings, but nobody is using it. There are too many falls in construction and general industry. All this permitting, organizing, and coordination can be done by the conductor of the symphony, and that person must know they are being evaluated on the success of the safety and health program and not on production. It is not hard to do it the right way, and the savings are enormous both in money and human suffering.

ENGINEERED SOLUTIONS

The engineered solution in safety is always important because when done correctly, it is the solution that eliminates the root cause of the hazard. There is nothing more important to the safetyman than making every effort to eliminate the hazard. There are problems and complications when the hazard is not eliminated and some administrative effort is not made to prevent injuries and illnesses. When the hazard is gone, there is no opportunity for a mistake that leads to an injury. There are hazardous chemicals that need evaluation, as well as the use of personal protective equipment. We should look at the possibility of the hazardous material being eliminated and replaced with another chemical. People tell me they never even thought about getting rid of a hazardous chemical when a safer option is available. The same concept applies to machine design that is protected with a barrier or guard. It might be possible to change the design of the machine to eliminate the hazard and the need for the barrier or guard. When recommending and acting for change, a specialist is needed, and it may be someone with an engineering or human factor specialty.

The safety professional must know when a specialty engineer is needed, such as a structural engineer, mechanical engineer, electrical engineer, or a civil engineer. I have seen too many safety professionals who try to do it all themselves. I never hesitate to recommend the right specialist to meet the needs of my clients. I recommend other specialists to design a guard or a fall-protection system and experts to install equipment like two-hand control systems, light curtains, and other presence-sensing devices. Coming up with engineering solutions to the safety and health problems is the primary responsibility of the safety professional. It is only after it is determined that it is impossible to engineer a solution to a problem that the safety professional begins to look at keeping humans away from the hazard. Protecting the human with personal protective equipment or posting warnings is the last resort.

Engineering a solution to a safety or health problem means eliminating the problem. It goes to the safety hierarchy that dictates that if a hazard can be eliminated, it should be eliminated, regardless of cost. It was decided early on at OSHA and at the National Institute of Occupational Safety and Health (NIOSH) that noise over ninety decibels should be engineered out of the environment if the engineering was feasible. The engineering was so expensive, and earplugs are so inexpensive. The industry rebelled against the engineering concept and the safety hierarchy when countless employers refused to spend on engineering control and just handed out earplugs. It is better to eliminate noise than cover it up, and covering it up creates some other safety problems, including communication problems where people with earplugs can't hear or talk to other people. There can also be infections from ear protection, and it's not always comfortable to wear.

In the end, it was determined that engineering solutions are not only feasible but also economically feasible, yet they are still not adopted by a society that wants to solve the problem with inexpensive earplugs. Still, these engineering solutions and the safety hierarchy are critical and important to progress, and personal protective equipment will never be as effective as eliminating the hazard. This same dilemma occurs with ergonomics. Many human factors and ergonomic hazards can be eliminated through engineering, but it is much cheaper and easier to try to remove the human from the dangerous situation or provide some device to attempt to mitigate the danger.

SAFETYMAN MUSINGS

NO GOOD DEED EVER GOES UNPUNISHED

"DIRTY DEEDS DONE DIRT CHEAP," AN AC/DC SONG. AS A safetyman, I know there is a price for trying to do good deeds and good work. I think it has something to do with the yin and yang of the universe. The universe is in balance, and when you interfere with the balance by trying to help someone, there is a price to pay. When you walk around as a safetyman, people take notice, and they will punish you if you do the slightest unsafe thing. The other day, I was wearing my snow boots, and I didn't have laces tied, and I was criticized for this, even though there was no tripping hazard. People give me heat for driving a car with a six-speed transmission. This is a safe car with proven reliability, but because it is a small car, they think I am being hypocritical. I am being watched, and I must set the right example.

Being a safetyman reminds me of the prime directive on the *Star Trek* series. The crew could visit anywhere they wanted, but they were forbidden to affect the culture, the history of the planet, or the civilization they encountered. It is the same here. When you try to change the culture or help people, they wonder about your motives, and they resist. They may or may not change their behavior, but you can be sure you are creating backlash. I have noticed that when I befriend or try to help or give advice to an individual, it usually doesn't work out well for me. I am a friendly person, trying to be a good safetyman, and then the person thinks they know more than I do. I am not the sort of person who wants to take advantage of other people. My life, for the most part, has been happy and enriching and has kept me busy. I never need or want to take advantage of anyone. Being busy has always been important to me. My mother used to say, "Ask a busy person." I think she meant this as a way to describe how a busy person gets things done by managing their time. A busy person has no time for laziness or lapses. I was not sure what she meant, but it seemed to mean that she wanted me to be busy, and I have been busy almost every day of my life, including today. Being busy with my own mission, there was never time or reason to take advantage of others, and I avoid a lot of trouble. I have found everyone is not like me, and some of this quality is a characteristic of a safetyman. One area where the safetyman seems to be different from his fellows is in the area of putting off for tomorrow what can be accomplished today. I do it today.

BLAMING THE DEAD OR INJURED PEOPLE

It is easy to blame injured people or dead people because they are not here to defend themselves. It amounts to a one-sided investigation. Somebody falls in a hole that is not covered or from a building or structure. Protection from these falls is required by standards, customs, and practices. The safetyman thinks the focus should be on making certain nobody else will fall from the building or in the hole. What happens next is everybody blames the injured or dead person.

Nobody wants to take responsibility for the occurrence, so the best way to handle it is to find nobody to blame, except the person who is not here to defend themselves.

This may have happened to you. You trip over a cord, or you miss a curb and get hurt. The first thing you do is blame yourself. You do this because it is too hard and painful to blame the people who are responsible. You can't think of a way to blame the people who left the cord on the ground or who didn't warn about the curb. So you blame yourself, and nothing is changed. The cord stays where it is, and the curb is never marked, and more and more people are hurt, falling over the cord or missing the curb, and it goes on forever. People who are injured will say how clumsy they are and sometimes start apologizing right away when they haven't done anything wrong. The human behavior system is rigged to protect those who are really at fault. It's only engineers and safety professionals who anticipate and protect people from these dangers.

There is a natural limitation for what individuals can do to protect themselves. Humans only have eyes in their heads and must walk with our eyes pointed toward the horizon, so we don't bump into anything. That is self-preservation. I have asked people to try to walk around while looking at their feet, and they can't do it. I can see the light go on in their heads when they realize the way we walk in this world makes trip wires, drop-offs, and open holes a treacherous condition. If a tripping hazard or a hole is behind you, it is even more dangerous. You can't see it and forget it is there. When I was a child, we had a game where we got on our hands and knees behind another kid and had another "friend" gives them a little push. They would fall over the kneeling person and land right on their head.

HOW DOES BLAMING PROTECT PEOPLE?

Some things a safetyman learns are surprising. Superstition, voodoo, random numbers, and games of chance live outside the work of the safetyman. Early in my career, there were a lot of references to gambling about safety. Life is a gamble, and we can calculate the odds and determine if some activity had an "acceptable risk." To me, this is crazy because most accidents could be prevented with a little thought and by following safety rules. The hierarchy of safety involves engineering hazards out of the world and the workplace, and it works perfectly without any gambling. I can understand how somebody might want to try to calculate the odds of an event. However, when the chances of preventing an event are near 100 percent, there is not much to calculate. I also don't appreciate rewarding workers for not having injuries by giving them lottery tickets or other prizes for winning their health and safety for another day.

Worse than the gambling approach to safety are the voodoo and superstitions. There are talismans of safety like the Saint Christopher's medal or the plastic Jesus on the dashboards. I have also seen people blaming other people for their injuries and deaths to protect themselves. If you tell someone about a person who had lung cancer, the response is always, "Did he or she smoke?" If someone had a heart attack, "Were they fat?" People think that not smoking or eating too much protects from dying. I feel the same way about people who think there is magic food or an exercise routine that is going to keep them alive. I fully appreciate the value of eating right, and the same for exercise, but when it gets to the point of being magical, I think it works against the science and theory of my profession.

When James Gandolfini died of a heart attack, an acquaintance said, "He looked like he ate a lot of sandwiches." I think this is a mean response. Mr. Gandolfini was rotund, but so are a lot of other people. Why are we blaming him for his death? Do people believe they can protect themselves from accidents and injuries by some sort of magical superstition? I read somewhere the author Stephen King said he wrote horror stories so that horror doesn't happen to him and his family. The problem is if we believe this magic, then we are not as motivated to do anything to make our world safer. If all this bad stuff happens to other people, and it is their own fault, then we have nothing to worry about! We go about our lives and work and don't get hurt or die because that only happens to other people. When people trip and fall, they can blame themselves or other magical sources. Sometimes people laugh when they see people fall and even when they get hurt. Is it possible that we feel protected when other people get hurt or killed? Events like NASCAR, football, and bullfights make it seem possible that hazardous activities only happen to other people, and we are somehow safe and protected.

DUMPING

I have seen many instances of people and companies throwing gasoline, oil, paint, or pesticides on the ground. They think the rain will make it go away or that it is nothing to worry about. Now we know that agriculture may be contaminated with these chemicals. It is that shopping cart dilemma again. We know that some people don't care about the earth. It is a shame to see all the birds and animals living in rivers and streams that are filled with toxic material, and in my rural community, too many young children are drinking the well water with these same contaminants. Health is so hard.

PEOPLE THINK IT IS FUNNY WHEN OTHER PEOPLE GET HURT

You have seen the old films and cartoons where people get a hardy laugh when they see a person or animal get hurt. I am sure this is the chicken or the egg, but when did we start thinking it was funny when people get hurt? Is it those cartoons and movies, or did they make those movies because they knew we would think it was funny to watch? As a safetyman, I never think it is funny to watch people fall or get hurt. I have noticed that the first reaction when somebody trips and falls is people often laugh. It's an interesting phenomenon because it is not necessarily what you would expect of human behavior. There is no way of telling whether a person is seriously injured or not, and when laughing, the observer must be thinking nothing serious happened.

One time, we were playing baseball, and one of the players got hit right in the testicles with the ball, and everybody was laughing so hard, except the person who got hit. The laughing stopped when the player was in the hospital undergoing surgery. He lost one of his testicles, and the humor was completely lost. When I was a boy, we would do everything possible during our roughhouse play to make sure that when someone was falling, we would get the full enjoyment of the falling. We thought it was fun. Now I realize that it is only fun for the observer. I think of football. We watch players smash one another, and we think it is so much fun. Maybe this is the reason we watch football. There is a television show where viewers are encouraged to send in fun videos. It turns out these videos, at least the ones they show on television, are of people getting hurt or almost getting hurt, landing on their heads, falling off a bicycle, and getting hit by their children. Safety is so hard.

GETTING OLDER

The population is getting older, and baby boomers like me are living longer. We are living with the help of medical miracles that are being developed every day. Most of the people I know around my age have artificial joints, hips, shoulder, or knees or have had some surgeries that have kept them alive. People living longer and living more active lives is great, but it has some consequences. When one of us goes down, we go down hard, and we don't bounce up like a teenager anymore. I recently went skiing with a group of older people, and several of the expert-level skiers got serious injuries. It can be trouble when an expert older skier is skiing on terrain that they are not familiar with, along with other good skiers who ski the same terrain all the time. It only takes one false move, a rock, or any kind of a fall. There can also be faulty or different skiing equipment. Even the people who were on crutches were saying they would get fixed up at home with their surgeon and would be back on the slopes in no time. This is just with skiing, but this problem with older, active people is everywhere—biking, running, walking, weightlifting, and so many more activities. I have seen some oldsters on skateboards. Even the best safety efforts and programs are not going to be fully protective of an aging person taking chances with their life. Add alcohol, and you know what will happen. There is little or no effort to try to get these baby boomers to have a realistic view of their athletic potential. The cost of those injuries is going to be enormous. Living longer is wonderful for many, but it is costly for all.

NUCLEAR AND THE RADIOACTIVE TUMBLEWEED

I've been to several nuclear facilities. The most interesting experience was when I went to Rocky Flats in Colorado. Everyone seemed to be afraid of the place. I was still working for OSHA, and we were a part of a team, and my bosses were worried. They were afraid because we heard that the place was completely out of control. They had taken nuclear waste collected from the reactors all around the country, put it into fifty-five-gallon drums, and buried them at the Rocky Flats site. Not certain of their logic, but it may have been to get the radioactivity away from the big cities and send it to colorful Colorado. I understand it was also sent to Utah and Las Vegas, but the biggest mess was at Rocky Flats. Those fifty-five-gallon metal drums began to rust and leaked into the ground and the groundwater, leaking to the point where significant radioactivity could be detected on the property. The property is fenced, and there is security, but as soon as you enter, they give you a badge that senses the magnitude of radiation exposure. I was not on the property for more than a few minutes when my badge alerted me to high levels. I had the impression that people who were working at Rocky Flats had found the leaking radiation hopeless a long time ago. The real question about the entire excursion was the reason we were there, because this problem is never going to be solved. This leaking radiation is going to be around for thousands of years.

I have a souvenir from the trip. While we were driving on the property, I spied a tumbleweed. Being from Chicago, I rarely see tumbleweeds, and this was a big one. I pulled over and put what I called the radioactive tumbleweed in my rental car. To this day, I have a copyright on Radioactivetumbleweed.com. I wanted to take my tumbleweed home, and United Airlines would not let me carry it on the plane, so I checked it as luggage. I just put a tag on it and gave it to the baggage handler. When we got to Chicago, I went downstairs to baggage and waited until my radioactive tumbleweed came out on the belt. There were cheers from my crew at OSHA and other passengers who had seen my tumbleweed confiscated when I tried to bring it on the plane as a carry-on. I took my tumbleweed home and tested it for radiation, and sure enough, it has a small amount. It is hanging in my garage right now.

OSHA sent me to Northwestern University for radiation training, and I learned about radioactivity both ionizing and nonionizing, which is lasers. To this day, I ask questions when a doctor or a dentist wants to give me an x-ray. The predominant information I have taken from my experiences with radiation is to protect my thyroid at all costs. The effects it might have on your DNA are unknown. The security at these facilities sticks with me. They acted like they were hiding something. Once, when I investigated an accident at a nuclear power plant, I took a picture of something that my handlers didn't like, and they took my camera and smashed in on the spot. It was only a $200 camera, but it was destroyed. They promised me a new camera, and the next day via FedEx, they sent me a replacement camera. It made me wonder about that picture. I hope it wasn't Big Foot.

WHAT IS THAT SMELL?

It could be cancer! Those truck and automobile exhaust emissions. How close do you live to the road? How about the black diesel smoke from trucks? How about the smell coming from the factory or the farm or even the stuff we buy in stores? Recent studies show living under or even near power lines can cause cancer. People living under those lines have the most inexpensive homes. When I first met my wife, she was living near overhead power lines in a townhouse. I didn't like it, and I found out she never thought much about it. She didn't like the buzzing noises from the lines, especially when it rained. I had heard if you take a fluorescent lightbulb and hold it under power lines, it might illuminate. I tried it, and it worked. I think this demonstration helped me get my wife to move in with me.

Recently, there was a big stink about a company partially owned by a former governor leaking ethylene oxide. This is a big deal because ethylene oxide is a carcinogen, but I think this company was only special because the former governor was involved. What about all the other companies that are poisoning the air and the water? Not just the air, but they might also be poisoning us, our kids, and our grandkids. Business is business, and we hold these company owners and politicians in high esteem. These companies make so much money they have political power. Bad smells and sometimes even good smells warn us about chemical hazards. The next step is to use instruments to determine what dangers might exist and are associated with those smells. Not enough people are doing this important work.

SAFETYMEN MISTAKES

OUR SAFETY AND HEALTH LAWS ARE LAX. YOU CAN KILL PEOPLE
and never go to jail. I demonstrate this to my students. I tell them that
I could "accidentally" put them in a confined space without oxygen
and get away with it. I have the perfect crime. If I were to shoot one
of them or hurt one of them or even talk to one of them the wrong
way, I would be in real trouble and would probably go to jail. But if
somebody "accidentally" allowed some carbon monoxide to get into
the room and they all fall asleep, they would call it an accident. As
I said before, I don't like the word *accident* because it implies that
nothing could stop it or prevent it. Almost all the cases I investigate
could have been prevented. I think the word *accident* is a trick against
safety and health. If someone on a job or in a factory is killed, there
may be a big investigation, and it might make the papers, but after
the investigation, OSHA will give them a $7,000 fine. This fine can
be reduced with good lawyers, and the company may walk away with
a slap on the wrist. This scenario happens every day in every city
and state in America. These "accidental" deaths cost a few bucks.

If you think about it, a manager or a company who violates safety rules and allows a human to be exposed to imminent danger or some toxic chemical is committing a criminal act and should be punished appropriately. People maim and kill people every day, every hour, and get away with it.

Drunk or impaired driving is a good example. I am not talking about driving over the limit for alcohol; I am talking about driving while impaired at all. We know that a single drink can impair driving performance, but we let people drive while impaired legally. This is not to mention all the prescription drugs, marijuana, mushrooms, and who knows what else. But here, we have technology to the rescue. Uber, Lift, and self-driving cars solve many of these impaired-driving problems. Self-driving cars and computers don't drink, take drugs, or have any of the dangerous characteristics of humans.

I recently had a case that relates to the power and responsibility of a safety professional. If you are training people for safety and health or protecting people with inspections and investigations, you have to be contentious. In this case, a training instructor in a fall-protection course put students up in the air with harnesses and lanyards and let them fall to show them how good the fall-protection equipment was for protection. The problem was the fall-protection equipment was only as good as it was fitted and adjusted for each student. Even with the best fitting and adjustments, after the fall, with the student hanging in the harness, there is going to be some pain until the harness is released. In the best situations, such as on a stage in *Mary Poppins*, there is discomfort at the pressure point on the body. In this case, the straps that go around the thighs were too close to the genitals and were not properly adjusted. This male student had a severe testicular injury that is going to cause him problems for the rest of his life. A diligent safety professional would examine the adjustment of each student and make certain it is properly fitted. The same is true when your kids are strapped into the ride at the amusement park. They better be strapped in the right way.

CLEANUP IN AISLE NO. 2

Even the slightest injury will require cleanup. I had a hangnail on my flight to Phoenix, and I got blood over everything. I get paper cuts at least once a month that bleed for several minutes. In the grocery store or at the train station, I get cuts. On the way to a meeting, I reached inside my briefcase and was virtually impaled by a staple sticking straight out. It went into my finger a quarter inch and bled for an hour. This constant cutting and bleeding goes on everywhere all the time, and it's not just me. You can get dozens of diseases from this exposure, and some biologicals will stay around for days in clotting blood. Every time somebody is bleeding, there should be a cleanup, sterilization, and covering, and maybe even a tetanus shot if you haven't had one lately. There is a specific procedure that can be found in OSHA or by asking anyone in the medical field. The problem is that nobody is following these procedures, and bodily fluids are flowing freely without anybody reporting it.

One time, I got a little cut at my gym and went to the desk to get a Band-Aid. They were not even interested in where I got the cut, let alone going to the machine that had the blood on it. They didn't even have a Band-Aid. How can we humans be so messy? And why don't we do anything to try to keep one another healthy? These are the real questions we need to ask one another. Once, while teaching a class of flight attendants at the OSHA Training Institute, one of them told me that people pee on the seat all the time because the access to the bathroom is blocked by the carts sometimes for over an hour. Anyone who has flown knows you can't get by the beverage carts on the plane. I have seen people at the health club locker room pee in the shower and even in the steam room. Everyone knows public restrooms are full of terrible stuff. Sure, I know nobody wants to talk about the fact that we are so dirty and dangerous, but it is true and a real public health problem. I am told that many people get sick at hospitals, and I am not surprised. In my experience, the cleanup protocol at hospitals is not much better than anywhere else.

I FORGOT TO CHECK

Forgetting to check for safety and health items is a firing offense. Forgetting and forgiving is wildly out of control. Can you imagine a jet pilot telling us they sometimes forget to put up the flaps or lower the wheels, or a flight attendant saying they forgot to arm the doors? This would be so serious that it would make the national news, but it happens every day on countless construction sites and at industrial plants, and nobody even hears about it unless many people are injured or killed. People in these plants and sites have life-and-death responsibility yet very little accountability. We must devise checklists and systems for accountability for people who have this important responsibility. There are cases of someone operating a train or vehicle and forgetting to check on an important maintenance item or issue. There are people responsible for filling out checklists that just check it off without really ascertaining it is done. A world where people with responsibility have no accountability is a dangerous place.

WORK HARDENING

There is truth to the concept of work hardening, but it is often misused. The idea has been that if you are out there doing a task every day, like painting or lawn care, your body will become adapted and strengthened to the task. Professional athletes and others who earn their living through physical activity become work hardened. This is a good concept. People are hired and moved slowly into the workforce a little at a time. They are given enough time to become accustomed to the work and for their bodies to adjust. The result is fewer injuries. But this work hardening is no longer the reality of our work world. A farmer teaching his sons and daughters farming activities, or a painter passing his trade to his offspring will result in work hardening. But in today's world, where we use day laborers or hire someone from another trade or occupation, this doesn't work. Hiring a worker and telling them to start painting a house or do lawn work will turn out badly until that worker gets used to the physical labor involved in the task.

Many valued and accepted concepts nurture the idea of work hardening: apprenticeship programs in the unions, training programs for the military and for developing athletes, and doing exercise or stretching before beginning physical work. All these concepts are beneficial in preventing both accidents and injuries. Unfortunately, from my own experience of being given tasks with no training and no direction, and my experience as the safetyman, as good as the concept of work hardening is, few employers are practicing it. They treat the workers as if they are all the same physically. They think a worker who presents himself or herself for work has been preparing for that work physically for a long time. The truth is that the worker has not been practicing and may not even know what the physical tasks might be. The results are, at a minimum, pain and suffering and possibly a serious injury. There are people who head for the ski country in Colorado and don't take the time to adjust for altitude or don't take the time to do some physical exercise before heading for the slopes. This is the same problem we have in the workforce, and employers have an obligation to make sure that their workers can do the work.

PHYSICAL CAPABILITIES

We must hire workers with the physical capability to conduct their duties without injuries and illnesses. It is essential to keeping costs down and reducing pain and suffering. Each company and each manager must take the time to think about these issues. We can't and don't want to discriminate against any worker, but we still need to match the capacity of the worker with the demands of the job. It's a difficult process to get the right workers into the right jobs based on their physical capabilities. Fortunately, with new technology and equipment today, we have more adaptability. A senior can lift with the help of material-handling equipment. Disabled people can operate cranes and many other kinds of sophisticated equipment.

Safety professionals and human resources personnel must look at and examine the physical capabilities of workers and assure that they are consistent with job descriptions. Some people might develop disabilities from too much sitting, and others from too much standing. Some people have better vision and manual dexterity than others. We can't take the antiquated approach that people are all coming into the workforce in the same condition. The more we examine the history of a potential worker and the capabilities, the more certain we can be that the workers will function without damage. I once wrote a chapter for a book on workers' compensation on safety, and I was surprised that, in most cases, employers are not looking at history and physical capabilities of potential workers. There was also little consideration regarding human factors and ergonomic aspects of the work.

COMMON SENSE

I have realized that the safety problem is not a safety problem; it's a human problem. We don't think, and sometimes we don't care if something bad could happen to us. I have been told a least a million times that people need to use common sense, but I never see any common sense. Common sense is an enemy of those of us who are proponents of safety and health. The idea of common sense is contrary to the idea of creating and engineering a world that will accommodate the weaknesses of human behavior and the human body. In primitive times, it was thought that we would learn from our mistakes. The idea of getting burned by touching a hot stove teaches a child or an adult to be careful of hot stoves. The problem is we can't see or measure the heat of the hot stove. We can make a mistake and believe that the stove is not hot or that it is not hot enough to do permanent damage. Many people who rely on the so-called common sense will make one mistake, and it may result in a life-changing event. It is reasonable to think before a human receives a serious life-changing injury or illness that the human believes that such an event would never happen. Once it happens, it might be too late and too serious for recovery.

The problem exists in the area of personal protective equipment. Eye protection is a good example. People and workers who don't normally wear glasses or who wear glasses that are not fully protective of their eyes are resistant to wearing the protection. People take their eyes and their vision for granted. Most people don't understand the fragility of their eyes and vision. I explain to students and workers who are resistant to wearing eye protection that when using a tool or just being around particles, the particle might enter the eye and cause temporary blindness. A particle might get caught in the whites of the eye and scar the eye and have to be removed, or the particle might go into the center of the eye and enter into an opening to the brain, causing permanent damage to their vision. Many people do not understand the so-called common sense to appreciate the physiology and the importance of protective equipment. The same could be said for the physiology of lifting.

Studies conducted by NIOSH have shown the average human capacity to be quite limited. A well-maintained body may lift approximately fifty pounds if the weight is close to the body and evenly distributed. Most people think they are weightlifters and attempt to lift without a good understanding of the physics or the consequences. The result is tremendous losses in wages, pain and suffering, and the costs of treatment to our society. Common sense tells us we can save people from these life-changing events, the pain and suffering, and all the costs with planning and training and education, but none of it seems to work very well with our human population. The problem is the lack of common sense in humans.

Humans like to take chances and enjoy danger, and the younger ones who have never experienced pain and suffering think bad things will never happen to them. It is surprising people think common sense exists. It annoys me when the premise of some conversation about safety and health starts with someone saying that it is common sense that will solve the problem or protect the individual. Nothing could be further from the truth. I don't believe in common sense. People act like the world is being controlled by imaginary people who always do tasks in a safe and healthful manner.

Eliminating the concept of common sense helps me finds answers and solutions to prevent catastrophes. It is not okay to leave a manhole open and without a cover because common sense says people will be watching for open holes and avoid falling into a life-changing injury. This alleged commonsense approach fails to consider that people don't walk while looking at their feet, and they don't expect an open hole will be there in the first place. Add in that they might be distracted while talking or looking elsewhere, or they could have physical or mental limitations, and they might even be impaired with alcohol and drugs. An open hole means it is inevitable that someone will fall into the abyss and have their lives changed forever. People have perceptual limitations and problems. They might be having a bad day, or they might be ill with a cold or thinking about a fight at home. These mammals might be drinking to forget their problems. They might have been diagnosed with cancer. A certain percentage of these people might have a desire to end it all. Three times this year, my train has been stopped for hours because of suicide. Last year, one of the suicides turned out to be the safety director for the commuter train company who stood in front of the train.

If there was common sense, it would tell us we need to create a world that accommodates the things we know about human behavior. Create a world that understands how people really act and react. We do this because we don't want lives ruined and families ruined, but even if we take all the emotion out of the problem, we do it because it is a waste of money to have to deal with the medical costs of taking care of families who can no longer take care of themselves. The idea of common sense gets us nowhere! In the legal world, there is a concept in defense called "open and obvious." It goes against everything we know about human behavior and is like common sense. The idea is if we watch out, we can take care of ourselves. The idea is that everyone must be prudent and look out for their own safety and well-being, and if they don't, it is their own fault. This takes no consideration of the facts of human existence and behavior.

REGULATIONS

Writing down rules and regulations is important. There should be a playbook to use for communication, discussion, and progress. I have learned when rules are written down, they are given more credence and respect. One problem is interpretation. People reading rules of conduct could have a different interpretation and opinion about what it means. Safety people take the written stuff and compare it to what is happening in the world. The old saying "actions speak louder than words" applies in this situation. Written rules of any kind, including safety rules and safety programs, are nice, but they don't tell us what is really happening out there in the world or at the plant or construction site.

When arriving at a plant or construction site, the first thing the safety director does is to show me some book on a shelf and proudly point to their safety program. This is not impressive, especially since I know most of the programs are purchased from a third-party vendor and downloaded from the internet, and nobody has ever read them. I have found people will lie about reading this stuff. If you ask any person if they have read the manual, they will say yes, and then if you start the quiz, they get nervous. I am not sure why people think that because something is written down, they have done all they have to do. In safety and health, it's good to have a fine written safety and health program, but it is just a document that is on the shelf or in a drawer—a document that nobody has read or updated. It must be read, understood, and implemented.

It makes the job of the safety professional simple if they can find things in a written safety manual that are not implemented at the site. The safety professional uses the written material to compare with what is going on. This is what I do 90 percent of his time. The safety program is required and must comply with the standards, customs, and practices. Any missing elements can be identified and rectified. Because something is written down doesn't mean a thing unless it is being applied. The idea of written safety rules being the most important element of a safety program is a bad idea. Written materials must be reviewed and updated on a regular basis. We cling to old concepts and safety standards. Technology is changing at a rapid pace. Most of the OSHA standards go back to the 1960s when we lived in a different time. OSHA is still trying to enforce some of those same old standards today. There have been tremendous advancements in fall-protection and personal protective equipment, instruments, and presence-sensing devices. Relying on old written material or old equipment does not serve the purpose of safety and health.

SAFETY PAPER

The safety program is a piece of paper that gives us important rules to create order in our lives. The safety program must be more than a paper with rules. Some of those rules can help us survive and create order in our lives, but these rules must be updated on a regular basis. As a safetyman and a scientist, I believe these safety programs and rules were meant to be changed. The rules and the documents must change as the world changes and as society changes. I don't understand why this is so hard to instill in humans. When I was at OSHA, a regional administrator showed me the proper use of paper. He had a three-drawer method that has helped me to this day. The top drawer is stuff you must do right now. The second drawer is stuff you must do sometime in the future, and the bottom drawer is stuff you are never going to do, and you can throw that away right now. I know that people are very resistant to change, but we must understand that papers have to be updated on a regular basis. I know this is a simple concept, but it is a very important concept. If we would just update our papers, the world would change.

WORKING FOR THE MAN (EVERY NIGHT AND DAY)

I should have learned about working for the man during government service at OSHA, but I never felt the pressure of trying to please my superiors or worried about losing my job. It all changed when I took the safety director job at a large construction company. This company was one of the largest electrical contractors in the country at that time. I was responsible for the United Center, McCormick Place Annex, Terminal 5 at O'Hare, the tunnel between the airport and the strip in Las Vegas, the Boston Harbor Project, and about 325 other construction sites. I could see that people were scared for their jobs, and I saw people getting fired for many reasons. This feeling I had about people fearing for their lives at work still sticks with me and reminds me of how fortunate I am not to be working for someone else in my own business. I learned how it is that a worker will take a chance or try to hurry and maybe make a mistake. I can understand how a worker might try to do something different to get the job done faster or more easily. I think sometimes people don't realize that working for the man can result in accidents and injuries because they just don't work for the man. Working for the man means you have little independence, and your job and your family security depend on some other person. This is critical in under-standing why a worker will jump on a pile of trash to get it into the trash compacter or climb up on a rickety ladder or scaffold to get the job done. These workers have no power, and if they get labeled as complainers, they are gone. A thousand times, I have had a manager tell me that the injured or killed worker should have just refused to do the job or complained to some higher power about the working conditions, tools, or equipment.

The trouble is that when you are working for the man, it's not realistic to think that interfering with production or complaining about safety is going to do you any good. The question of working for the man becomes a central issue in safety. In the old days, the worker could call their union steward and file a grievance or a complaint without fear of being fired, but today, things are different. The managers are more powerful than ever, and complainers are not usually valued as a resource. I believe the secret for improvement is to start to value complainers. I find that complainers are always telling me something I need to know, and that is valuable to me even if I don't like to hear it. Promote those who have a natural proclivity toward wanting to make things better. In my life, I have liked to complain and try to make things work better, and most of the time, I just get criticized.

Recently, when I was working on standards, I was complaining that the computer system the organization was using wasn't giving me the information I needed to do my job. Instead of appreciating my criticism, the manager got defensive and angry. Being an independent safetyman, I was able to walk away and go somewhere else where my ideas might be more appreciated. The problem is that almost everybody I know doesn't have that kind of luxury. Good employers and managers must learn to value people and reward people who are interested in improving operations, and unfortunately, this is rarely the case. I do think there are some places, mostly new companies like Google and Microsoft, that have figured this stuff out and appreciate creative thinking and complaining. This complaining concept needs a lot more emphasis.

HIRING THE RIGHT PEOPLE

I had an interesting experience when working at OSHA. I was working for the man, and the man was treating me like I was "the man," and everybody knew it. Believe me, this was the only time this ever happened to me in my entire career, and it was a short window. I was promoted and made the head of training. My job was being supported by an administrator who allowed me to have an assistant, and there was a big push at that time for upward mobility for women. There was a young woman who had received some attention for an upward mobility position, and my boss suggested to me I could hire her as my assistant. Before long, I had a bigger office and new desks, and there was Peggy doing all the clerical work that I no longer needed to do. This was great, and I was a good boss, and I supported Peggy in every way, even putting her up for raises and awards for her fine clerical work.

This went on for about two years, and everything was fine until my boss got transferred to Washington, DC, and in came a new guy who didn't see me as an asset and who, in fact, saw me as a big threat. He surrounded himself with yes-men, and I was loyal to the other guy. I got sent to inspect grain elevators in the winter, and Peggy, who had little experience or education, was moved into my position and then later into management. Before long, she was a big honcho. This reversal of circumstances taught me a lesson. When you are given gifts or power from some other person, it's not really something that you have earned yourself, and it can be taken away in a flash. For the record, Peggy became a top manager at OSHA, and as far as I know, she did a great job. I do remember, however, one time one of my clients was in trouble with OSHA, and I went to see her in her office to try to help my client avoid significant fines. She made me wait an hour before seeing me and treated me without any respect. I never again represented a client at OSHA. I had them hire an attorney.

THE NEW GUY

The new guy that came in to replace my boss and who moved me out was called the Captain by just about everybody. I was never sure what was wrong with him, but I knew he didn't like me. In some ways, he helped me get away from the bureaucracy, and his bad treatment pushed me ahead for better things. Sometimes I appreciate the people who helped me in my career more than they might deserve because it was the people who didn't help and didn't like me that had a bigger effect on my motivation to move on and be more independent. I tell younger people who work for the man that they should always be looking for their own job and their own company and to be independent, because somebody might just come in and tip over the apple cart.

BEING NEIGHBORLY

Those are words from long ago. Being neighborly was a regular way of life when I was a little boy. The neighborhood was our oyster. We could go anywhere without fear, and everyone we met seemed nice. My mother (Phyllis) knew everybody and was friendly with everyone in our building and the neighborhood. When we went shopping, there was a butcher shop, a vegetable shop, a flower store, and a candy store, and each place had an owner who knew my mom and was nice and friendly. My dad (Syd) bowled at the local bowling alley, and sometimes I went with him, and everybody seemed happy and neighborly. I remember taking the streetcar with my mom when we needed to go outside the neighborhood. That streetcar with the overhead wires and cane seats was so clean and friendly and neighborly. It is hard for people today to imagine how wonderful it was not to live in fear. No fear of gangs, guns, drugs, or anything. Life was but a dream, and everybody was neighborly. There must be a million reasons why it changed the way it has, but the concept of being neighborly has gone away.

In the safetyman world, however, somehow, people think that being neighborly still exists when they want something done. When a neighbor needs or wants to move a couch or push a car, these are the neighborly things that can end up bad from a safetyman's point of view. I do not respond well to neighborly requests unless they are for safety advice; I am an expert and have the answers and the proper solutions. If you want my advice about moving heavy things like chairs, tables, cabinets, sofas, or whatever, I will tell you to get an expert mover who has the equipment and experience to do it right. To me, this is like trying to pull a tree out of the ground, in that nobody really knows the forces that might be involved in this operation. Got an air conditioner that needs to be placed or removed from a window? Don't call me. I know you think it is wrong, and perhaps you think I am a jerk for not helping, but I am telling you it just makes sense.

There is the one about the flat tire on the road or the highway. I can't tell you how many times I have seen some neighborly person pull over to help someone in need. Then I see them on the shoulder or even in the roadway with the jack, and I think all it will take is a tired driver or maybe even a good driver who swerves a little bit, and that is the end of the neighborly person. Neighborly is not for me, and I think you should try to avoid it too. If you hurt your back, you are hurting yourself and also those who depend upon you. I am not saying we can't help one another out, because I believe in that. If you need safety help, I am there for you. If my neighbor is an electrician, he can help me with my new dryer. I am talking about helping when you don't know what you are doing. If my neighbor was a professional mover, I would love to use him or her to move my stuff, but I would also like to pay them for their service.

DRUG COMPANIES SELL DRUGS

Twenty-five years ago, I had this mysterious heart attack that I believe, but can't prove, was from a drug. It damaged my heart, and at first, it was scary and debilitating. When doctors look at my heart, they can see the damage, and some are surprised I have done so well. That's my own story. Today, almost every commercial on television is trying to sell us drugs, and if we ask for this drug from our doctors, they will give it to us. The fentanyl deaths in the news today tell a story about many doctors being paid to prescribe as much of the stuff and other opiates to patients as they could, and the more drugs they sold, the more money they were able to make. Don't get me wrong. Some drugs are miraculous. I have a friend whose wife has been alive for six years with lung cancer, and she would have been gone six years ago if not for this magnificent drug. The trouble is we, the public, have little knowledge of the science and testing of the drugs, and the drug companies are not eager to tell us about the side effects or potential hazards. This is just like the rest of the safety world. Some drugs help people, but for the most part, these drugs are being sold to make money. There is a big scandal and so much publicity about drug companies selling fentanyl and other painkillers to people who want them and need them, and to other people who act as middlemen to make the big bucks. The more they sell, the more they make.

YES-MEN

Not being a yes-man has caused me a lot of trouble in my life and career. I sometimes think it would have been easier for me if I had gotten along a little more. Even when I was a teacher, I could see the students who were nicer to me were getting better grades from the entire staff of teachers. It didn't seem fair. When I became a teacher, I had my grading system evaluate students based on a project they submitted. Getting along and being a yes-man have not been of interest to me, and I will turn down opportunities where I must be a yes-man. It is about maintaining my independence. I have seen many smart and tough people start to bend and then break into following somebody's orders. Many people are raised and taught to follow orders. I can remember when my parents sent me to Boy Scout Camp and I had a hard time following orders, especially from other kids who I never thought were looking after my best interests. Boy Scout Camp seemed like blindly following orders and doing what other kids were telling me to do. This contrasted with the words of my mother, who told me, "If one kid jumps off a cliff, are you going to jump off a cliff?" Following somebody else was not working for me. I just wanted to go my own way, and in those days, they called it marching to a different drummer. I certainly was a different drummer. They saw me as not cooperating, and I was passed over for every reward the camp had to offer. I couldn't wait to return home. I think my problem was I was always looking at the people in charge and wondering why they were in charge. I can remember thinking about the universe and wanting to be somewhere else.

A CHANGING WORLD

Our culture evolved from primates who seemed to care more for one another than we do in our modern society. These primates worked together to gather food and to protect the group from predators. Today, we must deal with predators. Some of these predators come from business. There can be big profits from selfish motives, and social media can be used to trick people into participating and doing things that are not in their best interest and even dangerous. Social media is full of tricks to get money from people as well as their votes. They will pretend to care about you to sell an elixir or drugs and make every effort to sell you things you don't need.

We need tools to prevent predators from taking advantage. The problem gets worse when you get old or infirmed. I have reached the age where every day, I get at least three telephone calls from people who want to take advantage. This is more evidence of the problem we have of not working together and caring for one another. Who can blame people from thinking that it all doesn't even make any difference? Nobody is doing anything about this, and it seems nobody cares. Some years ago, they told us that we could put our telephone numbers into some kind of do-not-call list. What a waste of time. They also told us that they were going to look after our air and water, but it was just a lie. Beware of anyone pretending to care. Caring is difficult, costly, and dangerous. I always read fiction. I thought if I were to write a book, it would be fiction. But now the world is so crazy I don't have to write fiction to tell a great and true story of what it is like to be a different kind of person—a safetyman in an increasingly dangerous and predatory world. A book of nonfiction is scarier than any fiction tale, and it is real. Just a continuing onslaught of predators. The safetyman wants a world where we help one another out of danger, where there is hope for change, and the world becomes a better and safer place. The safetyman wants our government to look after us and give special attention to the weak, sick, and old people who need the most help. We need a big change.

THE INTERNET

There is no stopping the internet now; at least I don't think so. So many people I know think it is obtrusive and taking away privacy, but I have never seen it that way. For the safetyman, the internet or the World Wide Web is a tool to get the information we need to solve problems and a way to disperse information quickly. We need to use the internet for good, not just for social media. The good and proper use of the internet could solve many of our problems. The internet is so good for gathering information. We can use videos to keep track of predators and all kinds of danger. To some extent, it does this already with all the cameras and videos; when something is happening in the world, we find out about it and fast. We could watch and report dumping, abuse, and violation of our laws. The internet and technology are already making a big difference in policing. We need to be watchful of any people, groups, or organizations that are private and don't want us watching. What are they doing that is so secret? I am optimistic that the use of technology and the internet can protect safety and health, but it can also do all kinds of good things to make our society more open and more representative of our people. The internet is a wonderful tool for safety. In the palm of your hand, you can have all the standards. You can have a checklist for inspection developed by the manufacturer of the products. You can have apps that measure noise, light, and even gas. There are still many opportunities to develop tools to help every consumer to use products safely. The internet can also warn us about the dangers in the weather. We can find out about traffic problems and avoid a multitude of other hazardous situations.

EMOTIONS

DEATH AND DYING

THIS IS A PART OF THE JOB THAT THE SAFETY PROFESSIONAL

does every day. Most people don't want to see death or injury, and they don't want to talk about it either. It needs to be talked about. So many people walk away from me when I want to talk about dying, and they get angry and tell me I am not being positive and that I should only think nice thoughts, but surely, they must see that without facing these issues, we can never improve the situation. I could write another book about it. People don't want to talk about it because it is too scary, and they can't face it even though there is no doubt that they are going to face it eventually. The job of a safety professional is to try to make sure that they face it sooner rather than later.

Some people believe in reincarnation or in an afterlife that is better than the life in this world, but we safety professionals are trying to keep people alive and healthy in this world every day. We try to get people to start talking more about the consequences of not believing in safety and health, and they need to see what it is like to live with a devastating injury or how people's lives come to an end when they are not properly protected. If we start talking about injuries and dying, we can make some progress in caring.

It's hard for us to understand terrorists who are willing to die and give up everything for their ideology or their cause. I don't think most Westerners can believe that there could be suicide bombers, except that there are suicide bombers. We know that they must be unhappy, and apparently, they believe they are going to a better place, especially the place with the virgins. It is interesting to me that that concept is somehow linked to sex with virgins. It's like the sex problem is so big here on earth that you can kill yourself and finally get the sex you want, which is apparently virgin sex, the kind that is all your own and that nobody else has had. When you add in the fact that most of these suicide bombers are men, there is really something to think about when it comes to sex and safety.

THE TERROR

So many of us live in a world where everything is good, and we have no expectation of terror except at a horror movie. Safetyman comes from a different place. I don't think I was looking for terror in my life. It came to me by my relatives getting sick and dying and people getting hurt. I thought the worst was always seeing people getting older, but seeing young people lose their quality of life is even worse. I know that I am acutely aware of these incidents. I have experienced everyday situations where I want to ask people why they are not practicing safe or healthful behavior. I want to ask them how they could be smoking cigarettes or taking illegal drugs. I also want to ask about aging, sickness, and dying. Nobody wants to talk about this stuff, and they always get angry, run away, or change the subject. Sometimes, people tell me that these subjects are taboo. I hadn't really thought much about things being taboo. It seems so primitive and tribal, and the stuff I want to talk about is real—real as can be. There are so many people who experience terror, probably all of us. This is a problem at the essence of safety and health. It is impossible to solve problems if we don't talk about them and approach them with reasonable solutions. I felt terror when my children were born. I feel terror when I have a new ailment or pain. I can immediately think of the worst things it could be.

In defense of myself, I know people with cancer, Alzheimer's, Parkinson's, heart disease, leukemia, and hepatitis. So why would it be so unusual for me to think that it could be one of those things? The reaction to my concern is that I am a hypochondriac. This hurts my spirit and my soul because I am being put down because of legitimate fears that I have. I think I am entitled to these fears and to express my fears without ridicule.

Recently, I was eating a sandwich and struck a hard object like a bone with my teeth. I had a lot of pain and then swelling in the gum over that tooth. I had once before bit into an alleged pitted olive and hit the pit and broke my tooth. It had to be extracted and implanted at great trouble and expense. I was certain that this was happening again, and I was unhappy about the situation. It turned out that it was a trauma but not a broken tooth.

Everybody treated me like I was a hypochondriac. My wife told me it was just a tooth, and even my dentist thought I was overacting. I don't think so. I was just facing the truth of my situation square in the face and not pretending that there was no terror in this world. This tendency to feel the terror and face the terror is very helpful and useful as a safetyman. I can understand the fears of other people, and I am relentless in search of means to avoid terror whenever possible. When I see an uninsulated electrical wire, I can see the electrocution damage from the entrance and exit wounds. One of my good friends has Parkinson's disease, and it is progressing. A year ago, he had what they call DBS or deep brain stimulation. This means he had electrodes placed in his brain. This helped with his tremors but adversely affected his speech. Recently, he went back in for more surgery to have one of the brain implants placed in a different location, and he had a seizure on the operating table. He is getting better now, but these things frighten the safetyman, and they are real.

People ask me the most about how I handle seeing all the injuries and death and how it affects me in my life. I don't think you can turn your back on the true nature of our dangerous world. It is nice to enjoy a day at the beach or with the family, and many of us have had great lives, but we know that bad things are out there and that there is no guarantee for tomorrow. Maybe this comes across more to a safety professional than other people in other occupations because we see the change that happens to the lives of so many people because of lack of safety and health programs and procedures, often just because they are at the wrong place at the wrong time. It could be the fault of aircraft maintenance that a plane crashes to the ground, but it is also the unfortunate passenger who happened upon that plane.

The relationship between the acts and failures of responsible parties and the chance of opportunity are some of the most interesting and fascinating parts of being or becoming a safety professional. Blaming and investigating all occur after the incident, and planning and preparation happen before the incident. The issue of what occurred at the time of the incident is fascinating. One person reaches into a machine; another takes apart a television. One person is so careful, and another is so reckless. Everyone has noticed the differences in people. From my experience, I believe that, overall, women are more careful than men. I have noticed that older people are more careful than younger people. I

307

think some cultures are more cautious. If I could return to my college experience, I can think of nothing that would be more rewarding than studying these differences in people and cultures to determine why people act in different ways.

INSECURITY

I studied psychology in school, but I didn't learn enough. I feel that I am an insecure person. I think we all are, to some extent. We were all pulled from our mother's womb, slapped on the rear, and set loose in a dangerous and hostile world. If that doesn't make you a little insecure, I don't know what would. Of course, people hide their insecurities in many ways. I hide my insecurities by pretending that I am confident, and in some ways, I am confident. Other people hide their insecurities by pretending they know everything. One of the smartest people in my life told me that being right is not important, and the price of being right might be too high. He suggested to me that it might be possible to give up on being right. This has something to do with maturity and humility, and I have found it to be some powerful stuff. This is not a self-help book, but if I were to write one, I think I would try to get people to stop trying to be right all the time. First, you might not be right, and second, there is no prize for being right. If you try to be right all the time with your spouse or your kids, even your friends and colleagues, you will pay a steep price, and it's just not worth it. A secure person doesn't have to be right. A mature person doesn't have to be right. A humble person doesn't have to be right. The problem is that we are surrounded by insecure people who not only have to be right but who also must be right all the time. I have found that there is manipulation involved with people who need to be right. They must use manipulation as a weapon of conquest. Recently, I was involved in a conversation about who has the best hamburger. I personally think In-N-Out in Arizona and California has the best hamburger. The people I was with think that Shake Shack has the best hamburger. A serious argument broke out about such a silly topic. Insecurity is a human characteristic that causes problems when trying to get people to cooperate and work together on something like safety and health efforts.

JEALOUSLY

They call this the big green monster. I never think that anyone is jealous of me. I am not a fancy person, not that good at sports, and not that rich. I don't care to have too many material possessions. It just never occurs to me that people might be jealous. The problem is that people are jealous of one another, and it causes many human problems and safety and health problems. I know a private detective who says that all murders are caused by sex or money, but in the end, it almost always involves jealousy. Jealousy can cause humans to do things outside of what would be considered normal. Jealousy is the enemy of developing cooperation and kindness.

LIES

There are so many things that people say to you that they believe are true but are not true at all. Buy and hold is one for stocks that seems to be wrong. There is the one about trusting your friends. Not so much. So many of the safety books and regular books talk about looking after your buddy. I think sometimes that buddies are not looking after each other at all. The buddy system just doesn't work. People won't be looking out for your safety and health; they will just want you to help them get some job done, and they often put you in the worst possible position, like the downside of the couch or sofa move. People think they can move themselves with a little help from their friends, but I can tell you this is a bad idea, even if you are young and strong. One mistake or bad judgment, and you might be paying for it for the rest of your life.

This concept comes into play with tandem lifting. This idea is that you can get one person on one side of a heavy object and another one on the other side and then lift the object and move it into the basement, the dumpster, or another location. This is such a bad idea from a safety point of view and even considering basic physics. The first problem is the cantilever that is about the balance point. Every tandem lift is like a teeter-totter because if one side gets higher than the other, then all the weight goes to one side, and the one person is holding the whole load. There are two situations that I see too often. One is when you are asked to help move the couch into the basement. I always ask if you want to be on the top side or the bottom side, because the top guy is just a guide, and the bottom side will be holding the entire load if it is possible. Another problem happens when one person in the tandem slips and drops the load. It could be from the slip, a bad handhold, or many other reasons, and this is where the other person gets hurt while trying to control the entire load from one side.

The National Institute for Occupational Safety and Health (NIOSH) has an equation that shows even under optimum conditions, an average person can lift only around fifty pounds, and that is only when the load is centered right next to the abdomen with what we call nonneutral postures, which basically means odd configurations of the body. I do a demonstration with a grocery bag inside the SUV, pretending it weighs fifty pounds. I have a person think about lifting the grocery bag from the end of the grate when it is near your body and then trying to lift it when it is in the back of the SUV. To avoid pain, people will slide the bag forward near their bodies.

The problem seems to be that we can't get away from our animal and predatory nature and become more caring. I am not a negative person. I have learned to live in the moment. I can appreciate the beauty of our world or planet and plenty of good things that exist, but in the big picture, it doesn't look all that encouraging. In the last few years, we have gone in the direction of caring even less for one another, our planet, and even our way of life. It seems that more of us, even our leaders, are trying less to look out for the benefit of future generations, and instead, they are grabbing what they can when they can. This might be the evolution of things that cannot be countered, but we need to talk about it. It's the hypocrisy and the lying. Everyone knows there is hypocrisy in business, in the church, in politics, and in our everyday life, but it seems like it is getting bad. We say we care about our kids, but we do nothing about future generations' natural resources. We say we want to make the world a better place, but we busy our lives with our own needs and pleasures. Instead of kindness and cooperation, we get drugs and gambling and schemes to steal money. More guns and more murder, yet we still think we are such good, kind, and wonderful people and that our nation and our beliefs are the best in the world.

If we look beyond the next scheme or the next fix and try to find some enlightenment and spirituality, we need to take an honest look at ourselves, our culture, our country, and our way of life and examine the good, bad, and ugly and make changes in our thinking and actions. We need to figure out how we can care for one another without danger and how to look at the truth of our situation without hypocrisy and bias. We need to think about the big picture in a practical way without the interference of cultural bias, magic, propaganda, or beliefs that we feel are true that might not be true at all.

On this earth, everybody wants to feel good. It's human nature and animal nature. How can I pleasure myself some more (porn it big) and feel good about the world and my life, and where everything is a fairy castle at Disneyland? Humans have raised themselves to a level where we can aspire to Camelot and fairy kingdoms. Disneyland was a great idea to make a lot of money, and last time I looked at their stock, it was doing pretty well, and it will continue to do well until such time as the earth is done, and then the roaches will be in charge of the fairy castle. Everybody wants to live in fairyland, and this is not going to change. This fact that nobody wants to hear or talk about the truth of these safety and health matters is a big problem. We want for ourselves, and we compete against other people. We live in a world of fantasy and war. Our kids spend hours playing or watching a game of winning, war, and killing, and this is not good. It is time to look at what is wrong with this world, so maybe we can at least think about it.

LYING

Lying is an important means of survival. It is important because many people don't want to hear the truth or know the truth because it can be a painful and depressing truth. Lying starts at an early age when we find out that when we scream and cry, we don't get that much attention, and when we keep doing it, people get tired of it and go away. Lying is a way of life in America. We never seem to tell the whole truth. We embellish the truth and try to put everything in the best possible light and look at the world with those rose-colored glasses. It starts early with Santa Claus and the tooth fairy and moves on to how smart we all are and how beautiful we are or could be.

I have known people who have achieved amazing success with the use of lying. They have had business success and even romantic success. Lying for companionship and sex is practically a way of life today. We have websites where people publish old photographs or even photographs of other beautiful people to get dates. I am told that most people lie or at least embellish on their résumés. How can we trust one another when we almost never get to the truth? It seems to me that if you are looking for the truth, you will be disappointed most of the time. The problem with courtship is our thinking that if you tell women and men the truth, they will probably move on to some other man or woman who will lie to them. It's the old self-fulfilling prophecy. Look for the right guy or woman or the right solution, and you will be attracted to the shiniest object, not necessarily the most truthful. Everybody wants a prince charming or a princess or the easiest solution to whatever problem they have, and since actual relationships and solutions to problems are more difficult, we tend to hide the hard work of the truth. No wonder there are so many failed marriages and relationships. It seems that the truth is the enemy of any successful romance.

I want to live in a world of empirical truth. I just had a guy from PayPal tell me that I would only have to wait two minutes to talk to a specialist, and now it is forty-five minutes later, and I am still on hold. There is nobody that will tell you the truth anymore. I am sure that guy has his own problems, but there is no reason for him to lie to me like this, and it just makes me angry at him and his company.

LYING TO OUR CHILDREN

I don't like when we lie to our children, even about Santa Claus and the tooth fairy. It starts our kids on a road where lies can be real, and it only leads to disappointment when they find out that the lies are not true. Oh, I know how cute the kids are on Christmas morning and how sweet when they get money from the tooth fairy, but honestly, this is not really for the kids; it's for the adults who wish to be entertained by the kids. I know there may not be many who agree with this safetyman, but I think we would all be better off if we told our kids the truth about the world. They might not be as cute, but they will be better informed. It is a tough world, and it's a world where the priest or even Santa might molest you. The police officer might be crooked. Of course, nobody wants to raise their children to be depressed all the time, but we certainly could give them a little more of the truth, and if we do, they might grow up with a more serious attitude about solving our immense problems.

THE TRUTH

People are really questioning the truth, and there are more versions of the truth from different people. In the world of the safetyman, the truth is testing hypotheses, gathering evidence, putting together several versions of facts and commentary, and then testing all the versions until we find the most likely scenario. The truth is something that can be tested. There are two actual tests of the truth: (1) validity—that means we are testing the right thing we want to test. And then there is (2) reliability—that means if we apply the tests repeatedly, will we get the same results. We happen to live in the moment in history when there is little use of the scientific methodology outside the laboratory. People are questioning science and not taking the time and effort to apply statistical methodology to the variables to determine likely results. Without this kind of inquiry, we could not have the space program, tall buildings, aircraft, ships, or any of our wonders of modern society, so who wouldn't want to use these techniques to solve our problems? Many people believe the truth is what they are told by some trusted person that might be a relative, a friend, a politician, a priest, or a rabbi, but that is never going to be a substitute for pure science and statistical analysis. Hypotheses can be tested if we care to do the testing.

NOBODY CARES

How can it be that nobody cares? We have some history of caring. There have always been some people who care for other people. Our survival skills involve caring for one another and working in groups for the advancement and betterment of the group and society at large. We developed in groups where everybody needed to work together and contribute to the survival of the entire society. We come from the right place, but somehow, we ended up in this competitive world, where instead of working together to solve our mutual problems, we spend all our time trying to please ourselves.

Survival of the fittest is a place we have started with our animal nature. This is the opposite of needing one another and working together to survive in a cruel and dangerous world where our individual existence depends on winning and conquering the weaker animals and people. How can we work together and not try to win or conquer one another at the same time? It calls for values and leadership. From primitive times, the leader of the pack got the most food and access to the pleasures of life. We recognize the power of this force, the yearning to win and be better and stronger than your peers. I learned more from losing than from winning.

I became the safetyman because I was so affected by incidents and injuries and the devastating effects they had on people. I have always felt bad for people who were suffering. The safetyman idea came from working in the field and coming to the conclusion that to have a better place, each man and woman needed to be a safetyman. A person like me needs to take action to represent safety and health, guard the machine, lock out the electrical, pick up the banana peel, and clean up some ice. Stand there until it is fixed or corrected because these are people's lives and their families that are in jeopardy. We might need to organize boxes, fix a damaged electrical cord, repair a ladder, or maybe wear safety glasses to protect our eyes. I believed people could become safetymen and safety women, and with a little encouragement, they could start improving the earth. The trouble is that as time went on, I began to realize people didn't want to become safety professionals; they didn't want to pick up the ice or wear the safety glasses. They acted (from my safetyman perception) in an illogical and unexpected way. They want free time, fun, and adventure. This problem seems counterintuitive to me in that this behavior is not in the best interest of the people's safety and health. It's not good for the family and not good for society.

I think one of the biggest things for me is the economics. It doesn't make sense to me to have someone get hurt when it could all be avoided with a little planning, a little caring, a little guarding, and a little protecting. I went about my life and my career trying to recruit an army of safety people to assist me in my effort to make the place better, and I found I was a captain without a crew. I have learned that people do whatever they want to do. They do not want to be told what to do by a safetyman. I wanted government regulations to make sure certain products and places were safe, and I learned people hate regulations. I find myself frustrated in trying to change the world, telling people they should put on their seat belts and their safety glasses. I always wonder what I can do to get their attention and have them listen to

good advice for their own benefit. I tell people that a hazard evaluation program is good for a guide, but it's the implemented efforts to
remove hazards that are important. There are four things to look for
in eliminating danger:

1. Specific rules
2. Communication of the rules
3. Monitoring of the rules
4. Enforcement of the rules

If you don't have those four things, you have done nothing!

There is a real possibility today that nobody cares about your
safety and health. Some people think the world is going to end soon,
so they don't worry about the consequences or other people or about
future generations. They are only thinking of themselves. People say
they care, but when you look at their actions, they might not really
care. When nobody is looking, they might throw toxic material on
the ground or into the toilet. The history of our culture is that many
people do not support a caring culture. Safety professionals stand out
against the rest of the culture. We need better-educated people working together for the greater good of individual safety and health and
the future of the planet. We need studies to find out why people don't
care about safety and the environment. We cannot ignore problems
right before our eyes. What does the average person do in a situation
where safety or health problems exist or can be avoided? We must
study humans as they exist so we can find solutions to these problems.
Most people use seat belts after thirty years of education, and most
people have stopped smoking cigarettes. This means that over time,
we can change human behavior, but it is not easy. If we start with the
wrong premise, such as common sense will solve the problem or that

the problem will go away or not affect us and our families, we cannot make the progress that is needed. How could we possibly change these attitudes? I am not sure of the answer, but I have noticed that in groups, people behave in a more civilized manner, especially when other people are watching. I have also noticed that when cameras are pointing at people, they are much more careful.

PROCRASTINATION

I don't have this problem. I don't know why. I just do it and move on. I never put anything off. This might have something to do with my obsessive-compulsive nature. I have never been diagnosed, but my daughter seems to have similar issues. It makes us great investigators and organizers. Since most of you are procrastinators, you may not understand what it is like to live in our world. I wake up in the morning and do the things that I must do, and I keep doing them until I am so tired that I must go to sleep. People like my daughter and I don't like weekends or holidays because those are downtimes where we can't be working and accomplishing our tasks. The last thing we want is to be lying on a beach, thinking about all the things we didn't get accomplished. A side benefit is that drinking, smoking, drugs, and other distractions keep us from our mission and duties. We are different. You must be different to be a safetyman. These same obsessive-compulsive characteristics make me a great safetyman and a productive and successful person. We are always on duty, and we are always looking for danger, violations, and changes for differences. We also annoy other people who are trying to have a good time. These characteristics make us a little antisocial. I am always investigating everything and everybody. This is the essence of being a safetyman. Since we are always working and focused on our task, all we must do is be in a position to charge for our time, and we will make some money. I was reminded yesterday by my daughter that I always told her that the world would tell her exactly what she is worth when she gets her paycheck. She works for the VA, which right now is on furlough because of the budget crisis. She is not getting a raise she was promised. She told me that there was a message in the lack of a raise, and I completely agree. She is a brilliant woman and psychologist, and they are not paying her near what she is worth. The important thing is that she knows it.

There is a downside to existing as a safetyman and not procrastinating and working at tasks all the time, and it involves the obsessive and compulsive part of the process. The problem is when you are working, studying, and investigating all the time, you make discoveries of tiny flaws in the world, with people, and even with material possessions that can drive you crazy. Everyone who knows me knows that I can't tolerate dings in my car or in any of my stuff. I also have trouble tolerating dings on people. My daughter is the same way. It takes us weeks to get over some small dent or scratch, and scars are intolerable. Somehow, I (we) have been able to have a happy, successful life being different from most people, but I can see as I get older that the higher power of the universe is making me accept a few dings and scars on my way.

THOU SHALL NOT TAKE ADVANTAGE OF WEAKER PEOPLE

The reason I say this is because everywhere I look, I see people trying to take advantage of other people. I get three calls a day where somebody is trying to trick me and take advantage of me. There should be a law that if you take advantage of weaker people, you get in trouble. It should be a crime to prey on the elderly and the poor. Our world is geared toward the freedom to take advantage of the opportunity. The rules are rigged against the weak and the poor. It reflects the survival of the fittest. In the natural world, the strong do take advantage of the weak, and the young take over from the old. They call it the circle of life, and they even have *The Lion King* to prove it. The trouble is that the kind of world that takes advantage of the weak is good only for the strong. We need a world where we are all taking care of one another, but we have a long way to go. Most people think of this as utopia and that we should just forget about it. A safetyman can't forget about it because there is too much pain and suffering that could be avoided with a little thinking and planning and a little kindness. The kind of change we need in this world is going to take a long time. If humans are still here and we still have a planet, that change is going to have to happen. It might come from a threat from intelligent life from other planets, or the children of future generations might come together and decide it is stupid to prey on one another. They will see that one day they will be old or challenged, and they will feel what it will be like to be in those shoes. I wonder why, as a country, we don't have more empathy for those who are not able to take care of themselves.

A KINDER WORLD

We say we want a kinder world, a kinder country, and a healthier environment. Nice to say this, but evidence of our effort often points the other way. Some may want a meaner world where they can hide behind fences and enjoy their great wealth and keep other people away from their treasure. It seems we are always at war. We can't get enough guns, and we don't want to take care of immigrants or the poor or the old. It's a problem to get to our higher angles when there is so much terrible stuff going around. We need to look deeper into the recipe for getting people to be kinder to one another and not so angry and fearful. Much of it is human nature, but we need to change our nature for survival. Why do we always have to win everything? There must be ways to calm everybody down. In the olden days, people were not so scared, but today there is a reason to fear every time we leave the house. The truth is that other people are trying to take advantage of us, our families, and even our culture. The world has changed, and now we must worry about foreigners and our neighbors. Our schools are not safe, and our theaters are not safe. We can't go to a concert or even to church and feel safe. It's no wonder that we are on guard every minute and that guns are falling off the shelves. We have the opposite of a kinder world and are going in the wrong direction. We could turn it around if we could just work together and trust one another. We must make people feel safe, and that is the fundamental job for the safetyman. It is such important work.

SAFETY MESS

This world is an amazing mess. It's complicated and scary, but at the same time, we are more prosperous and healthier than we have ever been in the past. We must take the good with the bad, but we also need to move in the right direction. We need to keep working to figure out ways we can be kinder and safer. We could spend a little more time on tasks that require cooperation rather than battling for winners and losers. We are going to have to praise our children when they work together and accept the differences in other kids rather than training them to be aggressors.

Sometimes I think that in my work, I can make a difference in this world. I try to relate to people from all over the world, and I have spent a lot of time trying to learn to speak the right language and send the right message. It is difficult to get other people to trust that I am looking after their life and quality of life. I work hard to try to get people to identify with me and my purpose as a safetyman. They have their own families and aspirations, and they want the finer things in life just like we all do. If I can show them how safety and health will increase production and help make the company more money, I can get them on board. I need to show them they can get a promotion or a bonus by riding along on the safety and health train. When this works, it is a cause for real celebration.

One time, I was hired for a job in New York. OSHA had referred them to me because the contract workers were dying on this project and two other projects. These were large residential projects and huge properties. One was in a place near Danbury, Connecticut, and there was also a place in West Palm Beach, Florida, and Aspen, Colorado. I had the good fortune to go to all three of these beautiful homes. The place outside Connecticut was being modified with a river and a waterfall. Trees were being imported for the property at a cost of over $1 million for each tree. Landscape architects would travel all over the country to find old trees for the property. They were purchased and transported for planting. There were almost a hundred South American workers on this project. Problems arose when several of the workers were killed while creating the waterfall on the river, where gigantic boulders were being lifted and placed like a puzzle or a pyramid. There were also numerous injuries when handling these large transplanted trees. It was my job to fix these problems outside Danbury and then move on to the other homes in Florida and Colorado.

The first problem was the language problem. I found that the bosses did not speak Spanish. I talked them into hiring some safety personnel who were bilingual and some interpreters to begin teaching safety principles in Spanish to all the workers. I pushed for the hiring of more bilingual managers and foremen. I complained when some of the bosses seemed to be uninterested in attending the training. There was plenty of work to do in teaching the proper use of equipment and coordination with engineering. I think I did a good job of preventing anyone from getting hurt or killed while I was there, but I never felt like I completely changed the culture.

Florida and Colorado were not much different, but there was not the same level of danger at those places. It was mostly regular renovation and construction. I learned a great deal from these projects, and most of all, I learned about the importance of language and cultural differences. A language barrier significantly adds to the safety and health program, and cultural differences can be a problem when workers come from a culture that doesn't put emphasis on the value of human life or the need for safety and health standards. One time, I taught a safety class with an interpreter to workers from Vietnam. I had some slides of the danger of cleaning up minefields. The idea was to teach the students not to approach danger and to ask for assistance. After the class, one of the leaders of the group came up to me and drew a picture of a minefield. She pointed to one of the mines and said, "Metal is money!"

SPEAKING OUT

Speaking out can get you in trouble with people who don't agree with you. This is truer today than ever before, and I am concerned that everyone will not agree with my opinions. I just have to have the courage to say that I am a safetyman, and I am the one who has lived this life and been to these places, and on that basis, I am entitled to have an opinion about these conditions. I can only tell the reader about what I have seen and what I have learned in my life and career. The reader's experience could be very different from my experience. I have never been afraid to be wrong, and when I don't understand something or if something seems wrong, I have never been shy to speak up, and this can sometimes get me in trouble. Being independent really helps me be strong for the principles I believe in and wish to support. For the many years, I have participated in safety, and in the development of safety standards, I have always had to battle people with different ideas and perspectives, and I have had to compromise to make progress. As I got older and wiser, I began to realize that compromise was necessary for advancement, but it has never stopped me from speaking out when I thought something was wrong or against the principles of safety. I have been to many meetings and conferences where I thought people were not making safety the highest priority, and it used to make me angry. Now I try to take it in stride and just keep speaking out, and I have found that I can accomplish a great deal.

DOWN AND DIRTY

I have seen a lot, and I should talk about it! First, it was fingers lost in machines at the sausage company. In an investigation a week ago, a woman got her hair caught in a machine, and it tore her scalp from her head. I remember a man who had his arm torn off, and I know he committed suicide. I inspected carnival rides at amusement parks where conditions were so bad that I feared for the loss of life and the destruction of families trying to enjoy a day together. I wanted to tell those people not to let their children go on those rides. I went to the slaughterhouses and saw how animals are killed. I think about the people who are employed to kill those animals. Day in and day out, death and misery. It takes a toll. At so many different companies, I inspected and investigated machines that will kill you instantly if you are in the wrong place or if something goes wrong. At one automobile-manufacturing facility, a radio-activated crane failed to respond to the signals, and a worker was maimed. Then there were the steel mills, where a few times I almost got killed myself. There is the job of approaching and tapping the blast furnace and the job of being a cobbler, where you move the molten steel around while it is still red-hot. I have investigated deaths at trash compactors, from, forklifts and with so many other pieces of equipment. I have been to many places that remain frozen in time, where over the years, the safety and health programs have hardly advanced at all—places like the steel mills, the railroads, and the grain elevators where explosions continue to occur. The loads continue to fall from the cranes, and the cranes continue to fall on people. I have seen death in tunnels and in confined spaces where there were toxic chemicals or not enough air. There were explosions at glue factories, fireworks facilities, chemical plants, and the refineries. When I first went to work at OSHA, I was asked to investigate a fatality at a glue factory. They were using acetone for the glue, and one of the machines blew up and killed a man. I went there and found no instruments to measure the explosive level of acetone. I remember writing in my report these exact words: "This place could blow up again at any minute." Two weeks later, it did blow up. I wished there was more that could be done at OSHA to shut down facilities before they blow up. Unfortunately, OSHA does not have the authority to legally shut down a facility without getting a warrant and going through a lengthy legal process. By that time, it can be too late.

LEADERS

Our leaders sometimes get to be our leaders by taking money from organizations that expect these leaders to do their bidding. Then when they enter public service, they are expected to follow the other political leaders in their various parties. We must have a new set of rules for our behavior in this country. We need more leaders and fewer followers. We need to have more people who think for themselves. We need cameras and security, but also, our objective has to be more caring. Abnormal, uncaring behavior has to be reported and monitored. In my life, I have seen our leaders and our institutions teach our children how to be followers at school, at church, in organizations, at work, and in life. We must teach our children to lead with confidence. This is a big change that is necessary in our world.

My parents taught me and my brothers to think on our own and speak up for ourselves. When I got in trouble at school, my mother's first thought was maybe it was the teacher's fault. I remember in high school in the summer, I used to cut out of school once in a while to go to a baseball game, and when I was missing from class, they called my mother into the principal's office for me to get punished. My mother said to the principal that she could understand how I might want to escape the school in the summer for a day of two to see baseball, and if my grades were okay, my punishment should be minor. I think I got my toughness and my willingness to stand up to authority from my mother.

BIGGEST INFLUENCES

JOHN LENNON

IMAGINE—THAT'S WHAT WE MUST DO! WE SIMPLY MUST IMAG-
ine how much better our world would be if we didn't take advantage of one another and if we didn't need to live in fear of one another. We would need to establish some sort of trust. Trust comes
from unpeeling the layers of an onion of fear and pain. They say
there is always hope from future generations. Right now, we have
the millennials, who seem immersed in technology, but they are
also more generous and forgiving and more understanding of individual differences. Maybe we can teach people to care. Maybe the
uncaring people of my generation will die off, and the new generation will care and do something about these great problems. As
a safetyman, I have learned to care for a variety of reasons. There
is always hope.

SHUT 'EM DOWN

Shut 'Em Down is a true legend at OSHA. He got his name because when he was a state inspector, he shut down the largest building in Chicago for safety violations. Nobody had ever had the nerve to do something like that, but he had nerves of steel. He was a great construction guy. I met him at OSHA. I could tell he was somebody important because he got the royal treatment. After a few months, I was sent out into the field of construction with Shut 'Em Down, and it was amazing how much he knew about construction equipment and safety violations. He took me into the deep tunnel project in Chicago that had only been going for a few years. It was dangerous in the deep tunnel, several miles deep under the city. He was the boss down there, and I could see everybody from the general contractor to the workers respected him. He was my first example of being strong and demanding respect as a safetyman. Shut 'Em Down was also concerned about my career. He talked to me about eventually going into consulting. He used to come over to my house and play with my kids and fix things that were broken in my old house and tell me the old stories in safety. Shut 'Em Down was an inspiration for me for many years, and although he has passed on from this world, he remains an inspiration to this day.

LIFE OF A SAFETYMAN

MY LIFE TURNED OUT TO BE THE LIFE OF A SAFETYMAN. I SAW
so much trouble in the world for people who had been injured or killed.
It's my job as the safetyman to tell you what I have seen. It is a story
that needs to be told. Rich and poor can be natural enemies. Often,
it is the poor who do the dirty and dangerous jobs that call for the
assistance of the safetyman. Being a safetyman means being at the
balance point of the scale of fairness between the powerful and the
poor. We are there to help the poor, but we are there at the whim of
the powerful forces that control our economy and government. When
you represent OSHA as a safetyman, you represent a third party, the
government, and both the rich and the poor distrust the government.

The famous joke about OSHA is the inspector saying to the
employer, "I am here to help you!" Managers have told me OSHA
stands for "Our Savior Has Arrived." Sometimes jokes tell a true
story, and in this case, the joke tells the story of how the public feels
about a government agency that was designed to protect workers. The
idea of OSHA was to make certain that each worker in factories and

construction sites in America would go home to their families without injury or death. There is a lot of dangerous and poisonous stuff out there, including more and more chemicals, like asbestos, hydrogen sulfide, lead, mercury, and ethylene oxide. We all have reason to be worried. Despite these fears, nobody makes changes.

One of my daughters, who was in high school in the fall, came home and told me she was picking strawberries. Being a safetyman, I thought about pesticides that might be on those berries. I asked what they were using and eventually got the material safety data sheet (MSDS). The pesticide had cancer-causing agents and mutagens, which means potential genetic damage and deformed grandchildren. You might imagine that I refused to let my daughter pick strawberries without the proper personal protective equipment.

Recently, they discovered that using Roundup fertilizer, which we have all been using for years without a respirator, can give you cancer. I don't believe I have cancer at this time, but every time I used Roundup, I got a headache. There is a furniture stripper you can buy anywhere, and you should be sure to read that label and try to follow the instructions. Somebody is supposed to be looking out for us. Everybody knows the story of the lead in the water in Michigan. The local government was involved in a conspiracy to keep the information out of the hands of those poor families that were drinking the water. I am not sure how we can make the government take care of us and look after us, but I know that what we have now is the proverbial fox guarding the chicken coop, and some people think it is okay. They have put the coal mine owners in charge of safety at the coal mines. Politicians seem to be doing everything they can think of to eliminate regulations—regulations designed to keep us healthy, alive, and safe. You would think we would be doing more to protect our lives, if not the future of the earth.

Discussing the problems is important. We must care for ourselves, other people, and the environment. There will always be people who do care, and we need more of them, and they need more power. People who do care must find their way into powerful positions. If we could get all those positive thinkers and the other people to look out for their own interests, we would make progress. Everyone needs to understand science and the reality of the problems we have in our world and in our society in order to develop a good plan. These plans should be coming from good leadership in a Congress that is looking after the best interests of our society and our country. Many of our current leaders are just looking after their own interests. Attempts to provide leadership on important issues have come from people directly affected by the lack of action or direction. We have the people who lost their children when somebody targeted children in the schools. As far as I know, we don't have enough powerful advocates for a safe world and workplace. I think Vice President Al Gore showed that one person can do a lot working for our environment. We need to band together and do something before it is too late.

I became a safetyman because I care too much about the protections that are supposed to be guarded by our Constitution. I can tell you our Constitution and all the rules and regulations in the world mean nothing if they are not enforced. How many times have I been told freedom isn't free? I can tell you that life, health, safety, and the ability to pursue happiness aren't free either. It's going to take a lot of work to get it done. Sometimes it seems nobody cares except me. My best effort is to reach out and tell the truth and inspire others to do the same. I am worried about so many things! There are so many promises that have not been kept, and the world is getting more and more hostile. Being a safetyman makes me find so many things to worry about. I know I can't be the only one who understands these promises that are not being kept, but where are you all? "You could say I'm just a dreamer, but I'm not the only one."

When I first went to OSHA, there were more than sixty compliance officers, and most of them were new to the program. The administrators had us read the OSHA Act repeatedly, and then we were required to read the standards from cover to cover. This was so boring, but it turned out to be so helpful for the rest of my career to know the standards. This was the pain that led to a gain. We did have government cars, and after a few months, we were given the freedom to explore the real world for safety and health. One summer on the news, we saw there were window washers falling from their scaffolds. There are specific rules for a secondary safety line that would have saved them. It was our job to clean up the window-washing industry and make certain those window washers did not fall to the ground. I cruised Lake Shore Drive every day and soon spotted window washers without secondary safety lines. I entered the building with my OSHA identification and conducted an inspection. Sometimes we would find deplorable conditions, no secondary safety lines, and no proper anchorage to hold the entire scaffolding. I issued citations to those employers and ended up in court against a giant window-washing company, representing OSHA. We won the case, and things got a whole lot better. As a result of that work, today, we almost never hear of a window washer falling to the ground. That is what it is all about.

I started attending meetings at various organizations to get the word out about OSHA and the safety and health effort. I found that I was welcomed at associations and unions and that many of these organizations wanted things to get better for their members. For the first time in my career, I could see that there were some supporters of what I was trying to accomplish. I had the opportunity to meet with the alderman and political leaders at those meetings, and it felt good to be doing something important in my life.

It was during this time that I learned to use computers in my work. I had taken typing in high school for two semesters, and now with the advent of computers, these typing skills were helpful. I had the opportunity to use the new computers to investigate issues of protection for myself and the other OSHA compliance officers. We were conducting inspections at fireworks factories that had already blown up, and there were fireworks that had yet to explode. I also had the opportunity to investigate the need for personal protective equipment for our inspectors. We at first did not have all that we needed. I also got to deal with cigarette smoking. In 1990, smoking in buildings was allowed. I had an idea to stop smoking in the buildings where we worked, and I am proud to say I had a role in allowing smoking in only designated areas.

I had so many adventures working for OSHA, and one of my favorites was when I got to work with the state inspectors. I spent three months working with the states in my region. I got to go out with many of the old-time inspectors, some of whom were seventy years old. I never learned so much in my life. They took me to the steel mills and taught me so much about what I needed to learn about the actual working practices in these large industries. I also had the opportunity to work in the grain elevators where they were having dust explosions. Spending time with these industries made me the safetyman I am today.

Near the end of my career at OSHA, I was given the opportunity to move to the OSHA National Training Institute in Des Plaines, Illinois. This was one of the best things that ever happened to me. This was the greatest time of my career, and I learned more than anywhere else, teaching those courses and meeting safety professionals from around the world. I told my bosses that I was looking to leave the government, but they never believed me. One day, a man who was in the back of one of my classrooms came up to me after class and said he was going to hire me. He asked me if we could go to my private office, and I told him we could go to my cubicle. He went to my corner cubicle, looked around, and said words that would change my life: "Frank, if you fail with me, you could always come back here!"

I knew he was right! I left OSHA for more opportunity, including a car and an expense account, and went to work for one of the largest construction companies in the country as their corporate safety director. As I mentioned previously, I oversaw 350 jobsites all over the country, including the United Center, Terminal 5 at O'Hare, McCormick Place expansion, and 347 others around the country. Most notable were the Big Dig in Boston and the tunnel between the airport and the strip in Las Vegas.

This was a big job, too big. I went all around the country, sometimes three cities in a single week, and visited jobsites, conducted training, and much more. I was obligated for one year to work for the big construction company. I was starting to meet people, including insurance company representatives who had construction clients. They told me they could use my services, and I knew I was in consulting. That is where I have been ever since.

WORLD TRADE CENTER

When I went to work for the big construction company, we had some of the biggest construction projects in the country. We were traveling all over the USA, promoting new safety programs and conducting training courses. We had a big corporate meeting at the World Trade Center. The first day before the meeting, I was conducting a training session on the third floor of the Vista Hotel. I will never forget that day because that was the day that terrorists blew up the parking lot under the Vista Hotel, and the whole building shook. I thought it was not only the end of my life but the end of the world. On that day, February 26, 1993, the World Trade Center was bombed for the first time. A bomb built in Jersey City was driven into an underground garage of the World Trade Center. The blast killed six people and injured 1,500 others. I can tell you the bomb blew up sometime before noon because the chefs had just come into our training room with two big pastramis. Next thing I know, as a safetyman, I am trying to lead people out of the building. I knew that it was my duty to leave everything behind in an emergency, so as I led thirty people down the stairs in darkness and dust (the dust was horrible), I learned to appreciate emergency lighting. I went outside, and it was very cold, and I had left my coat and my computer in the conference room. This turned out to be interesting because the next day, after one of the biggest events in history, I was able to get into the building to get my coat, computer, and briefcase. I went into the lobby that had been elegant and beautiful the night before, and now it was just a big hole. I couldn't believe my eyes. I climbed up to the third floor to retrieve my coat and computer and saw piles of concrete dust. The most amazing thing I saw, and maybe the most amazing thing I have ever seen, were those two big pastramis sitting there, still fresh and untouched but covered with dust. Life and death and food.

DON'T TAKE UNNECESSARY CHANCES —
BE AN INVESTIGATOR YOURSELF

I have a case where a machine was heating up beyond the redline for six hours. The machine had heated so much everyone knew it was a significant danger. They decided something was wrong with the gauge rather than with the machine. It is hard to imagine that they would blame the gauge. They investigated the machine with a flashlight. They opened the hatch, and the machine exploded. Everyone should know the right thing would be to shut down the machine and let it cool down. In a big organization where everything is pointing toward production, it is hard to get a leader to shut down the operation. There must be someone to make this call. This is what safety leadership is all about. This is the nature of the safety culture.

THE POWER OF POSITIVE THINKING

I have friends who talk about the power of positive thinking and trying to get everybody off their pity pot. I am all for positive thinking, but it doesn't work for everybody. I think that telling people who are suffering to think positively is not always the solution to their problems. In my experience as a safetyman, I have found it difficult to get people to think positively whey they have been subject to life-changing misfortune. Positive thinking is fine when it's not you in trouble. If you are not the one hurting, it's not a problem. If you are hurting, what you need is some comfort and understanding. Depression and addiction are big problems in our society, and when you look around, there are so many people in trouble who need help, and they turn to drugs or crime to get them through the day. Positive thinking does not always help us. It's a good idea to try to put the best spin on a negative situation, but we need to be careful not to shame people who have real problems.

Those rose-colored glasses are fine when everything is great. When asked how you feel, you have never told the truth or that you are worried about something. You keep it inside. This is an enormous human problem of not expressing our true feelings. Positive thinking is overrated because it keeps us from looking at the problems in our society and our world. Pretending everything is great is fine, but it doesn't make our problems disappear. Sometimes it may make them even worse. There is much mental illness in our world. Nobody wants to talk about problems. Everyone just wants to pretend their life is like life on some commercials on television or on social media. Everyone is smiling and happy, and it looks like everyone lives forever, stays young forever, and is perfect. So those of us who are not living the perfect lives feel like there is something wrong with us, and we are fearful of expressing ourselves because nobody wants to hear it.

SUPERCOLLIDER

The supercollider was an enormous project where they were going to split atoms by shooting them in a giant circle under the ground, and when the atoms collided with one another, they would smash into particles, and we would learn things about the secret world of subatomic particles. I got to see it. It wasn't too long after I saw the construction and conducted an inspection of miles of a tunnel dug into the ground that the entire project was canceled for lack of funds. I wondered what happened to that excavation and if it caved in or was covered up. The earth is still screaming about this violation, and it turns out it was for no reason. Tunneling was one of my specialties because OSHA wanted me to be their tunnel coordinator. I was trained in tunneling, and I learned how dangerous it is to go underground. I learned that my life depended on air being forced down into the ground and structural work of roof supports and roof bolts. I was down under the earth while giant boring machines were cutting into the crust and blasting with dynamite. I can still remember being in Hawaii and inspecting the tunnel being constructed between Honolulu and the west side of the island, which they were calling T3, when they accidentally shot off some dynamite while we were still near the entrance to the tunnel. It was a close call. Inspecting tunnels in Hawaii, Chicago, Boston, and at the supercollider gave me great respect for the danger of working underground. It was the same in the coal mines in West Virginia. There was always the possibility of something going wrong. The air was thin and dusty, and I was always happy to get out of there.

HOW STATISTICS CAN LIE

A billion is hard to get your mind around, and now we are dealing in even bigger numbers. We become overwhelmed by statistics, and everyone knows that statistics don't always tell a true story, and they can be manipulated. It would be easy to just count the injuries, count up the lost workdays, multiply by a constant for size, and divide by the number of hours worked. Then you could have a simple means of comparison between one company and another, one industry and another, and one location and another. Only one problem: nobody wants to do this. If you start counting these things, you will find the companies who have the injuries, health problems, and deaths. There is a lot of pressure to keep us from having this information. Unfortunately, this kind of information is what we need just to get started in reducing the losses.

Let's just say we wanted to decide where to eat fast food. Is it safer to eat at McDonald's or Burger King? It would be nice to have that kind of information. How about safety at Disneyland versus Universal Studios? Walmart or Menards? Overall, if we had statistics regarding where people were getting hurt, we could possibly avoid those places. I always wonder about these county fairs and these carnivals that just pop up in parking lots. I would like information and a comparison of accidents and injuries in order to decide if I will be going. I would like to know the injury records at resorts, and it would be especially interesting at places known to be dangerous, like the places that sell flying lessons or skydiving. When I see people zip-lining, it is hard for me to believe that they would do it if they knew about the lack of maintenance and history of injuries. Of course, there is the problem that they might not give you the right statistics, or they might even manipulate the statistics. Who are the statistics police anyway?

We should give the workers or the consumers a choice. Go where you could get maimed or killed or go somewhere where you know they are looking after you. Simple choice. Go to a steel mill or Disneyland. We will know where we can take ourselves and our family to a safe and healthful environment when we know the facts and the statistics of where people are getting injured or killed. If we had as much information about injuries and death as we have about sex offenders, we would be off and running into a safer and more healthful world. We can work for a company where our life and quality of life is in jeopardy or work where we know we will be safe. Nobody wants to get hit by a crane or end up dead in a trench or on a factory floor. Nobody wants to suffer from silicosis or asbestosis. The problem is, without good statistical information, we are unable to make the right choice. At a time when jobs are scarce, workers will take whatever jobs they can get, and they won't complain when things are not safe and healthful. Everybody is fine until something happens and someone gets hurt. Everybody will say it was an accident, but if we had the right information, the right data, we could have steered clear of the danger. This is a common problem in American industry because inside the human brain, there is a calculation that says things like, "I am never going to lose the cancer lottery," and "I am never going to lose the falling-from-a-height lottery." This is normal human thinking—that bad things are never going to happen to me.

SAFETYMAN IS DEBBIE DOWNER

We need to form support groups of caring communities of people interested in solutions rather than just throwing money around, winning the approval of popular culture. Without a support group, a safety professional can't do their job because the societal pressures against safety and health solutions are strong. Everybody wants to be a star, and nobody wants to be a villain. The safetyman or safety woman must play the role of the villain or Debbie Downer. It is not easy to be the villain and to constantly see people doing the wrong things and trying to correct their behavior. I have known a few people who can do this. It gets tiring, and it goes against human nature not to chill out or take a chill pill and stand firm and demand a change or correction that might save someone's life. It's interesting that when a fireman or a policeman arrives after an event and takes action, they are viewed as heroes, but when a safetyman or a safety woman arrives before something happens, they are often viewed as the villain. We might tell them that they should have a fire extinguisher in this area, or this electrical cord is frayed, and you can see that they are not happy, but they should be happy because somebody is trying to save their life and their quality of life. People do not want the safety professional spoiling their fun. Also, the safety professional is a human and wants to have fun too. The trouble is that the safety professional has an important job to do, and it involves avoiding the tendency of looking the other way. This is so difficult, and I have experienced the pain of trying to get somebody to cover a hole or move a tripping hazard, and they look at me like I am the meanest and most terrible person in the world.

QUALIFICATIONS

Certain key words are essential for the concept of safety and health. One of the most important words is the "competent person." These two words go a long way toward assuring that the safety and health of American workers are protected by somebody who knows what they are doing. The idea is that some educated safety and health professional will know about safety issues and have the authority and initiative to take appropriate action before the accident and injury occurs. These people are proactive and solve problems before something bad has a chance to happen. Often, this competent person will not be just one person but a group of specialists who have experience in a problem area and can take the problem-solving action. Too often, a generalist arrives with a single course or a "competent person" card to "bless" structures and equipment and even activities, without having had specific training and experience to know what to do about solving the problem. Safety and health generalists are very important but only to the extent that they know enough to find and hire the appropriate experts rather than try to solve the problem themselves. We must be careful to train our safetymen and safety women to understand that they don't have to have all the answers. They must be smart enough to find the people who have the answers. I remember reporters calling me and wanting to know why the cranes were falling into the streets of New York. The correct answer from me to those reporters was that those cranes would never have fallen into the streets if they had been inspected by properly trained engineering experts.

Real competent persons must get the help they need to solve difficult and complex problems. I noticed some companies don't want outside experts or consultants because they want to continue to get the good news from their in-house "competent person." A true competent person should at least have some independent credential or authority beyond the employer. A true competent person has experience with the specific problem and identifies solutions. This is one of the most important issues in safety because it involves the credibility of the safety professional and the profession. Having people running around with cards that say "competent person" is only going to make matters worse.

SAFETY SWAT TEAM

I have always liked the idea of the Safety SWAT Team. This is an operation like the Federal Aviation Administration investigates plane crashes. There would be a team in every area office that would be sent whenever any human was injured or killed, and they would investigate and make certain it will not happen again. There was an engineer once who worked at OSHA. He had a very simple but good idea to have OSHA actually investigate accidents and injuries with capable teams of engineers and other experts, which would result in studies and documentation of the problems that caused the incident. They would then inform the nation as to how the problem could be avoided in the future. There had been a serious precast concrete accident, and a whole structure had fallen. The team of engineers determined there was not enough lateral stability in the structure as it was being built, and they gave specific recommendations for how this problem could be avoided in the future. There was a technique called "lift-slab" technology that was virtually banned because of this incident, as the lift-slab technique was a major cause of injuries and deaths. Couldn't this idea be applied to every serious injury or catastrophe?

SAFETY POLITICS

OSHA, for political reasons, has been having trouble doing the job they were meant to do. They have been stymied from writing new standards and enforcing some standards. Corporate America does not want to see a compliance officer at their door, and they do not wish to pay fines and go to battle with the federal government. OSHA has become unpopular, and there is pressure for deregulation. A safety professional believes everyone should agree on providing safety and health programs, and policies are for the benefit of corporations as well as individuals. This negative attitude against OSHA could be turned around. OSHA could be a bastion of information on how companies can be productive and save money by being safe and healthful. Having OSHA or other federal government agencies involved in political issues is not productive and has not worked. There are other groups that can get involved with standards, such as ANSI and ASTM. Both do an excellent job of writing standards. ASTM writes standards for products and material, and ANSI writes standards for various occupations and activities. If politics could just get out of the way and standards writers could write updated and productive standards, we would all be better off. Let OSHA do the job of making certain that corporations and companies see the benefits of implementing standards and programs. Political football, which was evident when I was involved in trying to get an ergonomic standard passed, is unfortunate. Instead of everyone working together to solve the problem and reduce the costs of ergonomics, everybody was in their political corner, angling for victory. This problem almost destroyed the entire process. So many other standards are affected in the same way by political sentiment. We can do so much better when we show everyone that no matter what their political affiliation is, we can save them money and give them a better life with an ergonomic standard or a gun-safety program.

OSHA INSPECTION PROCESS

There have been many courses offered to employers on how they can survive an OSHA inspection. I would never recommend any of those courses. The OSHA inspection should be considered an inexpensive or even a free service from your federal government. It is so useful to have fresh eyes look over your operation and possibly save you from a fatality or a catastrophe. I tell my clients not to panic when the OSHA people arrive. They may find nothing at all or something small, but if they do find some big problem, they are doing you a tremendous service. Of course, the OSHA inspectors (compliance officers) are not perfect and have different amounts of training and experience, but they have all been trained to look for serious hazards. If you are a dogcatcher, you look for dogs, and if you are an OSHA person, you look for hazards and violations. They are not familiar with all trades.

An electrical inspector for the city has experience looking at electrical systems, and it is the same for every trade. OSHA inspectors tend to be generalists. They have special education and experience they receive at the OSHA National Training Institute, but you can't compare them to people who spend a lifetime learning to weld or put together machinery or electrical systems. When an OSHA compliance officer initially goes into the field, they will learn about hazards and equipment as job training. This job training takes years and perhaps decades before that inspector is seasoned in all the areas also called subparts. This means you may have an inspector not fully knowledgeable in your industry. The inspector may not have had experience in your specialty industry. Under these circumstances, the OSHA compliance officer may call another expert. At a minimum, the compliance officer will compare the safety equipment to an operation's manual or company procedures. They might just conduct only a partial inspection. It is possible the OSHA compliance officer may miss important hazards. I have heard companies who were visited by the OSHA say that OSHA approved their operation. Nothing could be further from the truth. OSHA people don't know everything, and most of them are not safety engineers. After a few years in the trenches, they will learn enough. They will keep getting better over time, and the OSHA compliance officers have improved over the years.

THE QUESTION OF ENFORCEMENT OF SAFETY RULES

Sometimes enforcement effort doesn't improve the safety and health culture. I remember when the OSHA officials believed companies needed to fear the federal government to do the right thing for safety and health. Enforcement and fines are the primary tools used to gain compliance in many industries, including the airlines, the FDA, consumer products, and many other places. The idea of enforcement of laws and rules is engrained in our culture. The problem is that when the rules are not clear, or the adjudication process is lengthy, or the punishments are too small, the system doesn't get the expected results. The institutions violating rules and regulations must fear the punishment.

There is a movement right now to avoid punishment and move toward rehabilitation. There is evidence that punishing people or corporations for violations of the law or safety and environmental rules doesn't solve problems. These institutions need the incentive to succeed, and I believe the profits and cost-saving of the effort should be enough to effect a great change.

So what can we do besides enforcement of safety or health rules or even laws to make the planet a safer place? The answer must be education and engineering. From a safety and health standpoint, we can gain profits when we minimize danger and remove hazards that will likely cause harm. The supermarket store solved the shopping cart problem by making people pay temporarily for the shopping cart. Other problems can be solved in this same way if we make people pay in advance for problems that need a proactive approach. It will be too expensive to fail to provide floormats or safety guards. Once people are educated on the savings and benefits that are achieved by acting on the front side, this problem can be solved without obtrusive enforcement action. The fact that humans and corporations must make money to survive can be turned into an advantage when we show them they can make more money by doing the right thing. This is what we need to do, and it will be far more effective than fines and penalties.

VOLUNTARY PROTECTION PROGRAM

VPP programs recognize the best safety and health programs. These companies get a flag and are exempt from general OSHA inspections. OSHA teams spent months at major chemical and refining corporations, examining elaborate safety programs, making sure everything was in order. A problem with the recognition effort was that the flags could not eliminate production pressure and human errors. The recognition effort, in my view, was not a successful way to improve the safety and health programs at these companies, and some of the statistical evidence seems to show they still had injuries and fatalities. It was discouraging to see recognition programs fail! There is nothing wrong with rewarding companies that are reaching forward in the safety and health culture, but it can become a political prize. It can never be used in lieu of a program that demonstrates corporations will become more productive and profitable using the best safety and health programs. When OSHA or some other group starts rewarding corporations for profitability through safety, we will move forward.

PLANNING FOR THE FUTURE

OSHA, NIOSH, and the National Safety Council can be clearing-houses for available experts to solve real and important problems that threaten the safety and health of people, workers, and entire communities. These organizations should be well funded to hire experts to anticipate serious issues we may face in the future and to resolve the root causes of incidents and injuries. These organizations should be a source of information and inspiration to society on the benefits of being proactive. Every year, we have clusters of injuries and illnesses and epidemics of weather, flu, and disease. We must have people and organizations dedicated to the collection of data and who alert us to peril. Countless lives will be saved when strategies are developed to prevent disasters. We need a Department of Organization and Planning to avert disasters. As it is now, the agencies that protect us from danger are scattered around the government. We need a centralized arm of our government that is proactive for our quality of life.

STIMULATION

Boredom while working at a task in a factory, at a construction site, or on the railroads causes injuries and fatalities. Humans need stimulation, and when the work isn't interesting or becomes routine, they try to make it more interesting. These repetitive jobs need more attention than they are getting. There have been techniques that when implemented have made a difference in the quality of work and the reduction of injuries. One of the best ways of eliminating boredom and injuries is job rotation. Many production lines have used job rotation to great advantage, but there is not enough of it being used in construction and other industries. Tower crane operators spend their entire work lives alone in the cab at the top of a crane, with no relief. Electrical repair workers do the same job repeatedly until they make a mistake. Air traffic control people get tired and bored too.

On the other hand, there are jobs that can be too exciting. There have been television shows and videos of the most dangerous jobs, like working on a crab boat in Alaska. I have been on a crab boat, and when I saw those pots being lowered and raised with live seafood, it was hard to avoid the moving ropes and heavy crab pots. It was very exciting but extremely dangerous. I could also see how it could end your life. One time, we brought up an octopus, and the knives started flying. Most jobs are not as dangerous as crabbing and not nearly as exciting. Operating machinery day after day, such as jackhammers, power presses, and saws, lifting loads, and connecting wiring is boring. Humans find a way to make it more interesting. I have seen so many shortcuts and tricks workers try just to have some stimulation during their day. As a safetyman, it is important to watch people and make sure they are getting enough stimulation and are not too bored or on drugs to get through the day.

This is just another of the human factors that doesn't get enough attention. I have seen some managers and foremen who were completely ambivalent to their subordinates getting hurt or killed, or maybe there have been a few who enjoyed seeing somebody get hurt or killed.

THE UNIVERSE

The universe is mystifying and incomprehensible. It's real, and we can see it with our eyes and with the Hubble telescope. The universe and infinity identify the things we don't understand and open the possibility for the future. I remember the movie *The Day the Earth Stood Still*. An alien from another planet told us we must start caring and stop killing one another. This is a great concept, if only the aliens would come and help us stop hurting ourselves. But until the aliens come, we are going to have to figure out a way to be kinder and more considerate to one another. We need a safetyman. Making us do the right thing for our planet and people is the domain of the safety professional because we know how difficult it is to get people to change their behavior and attitudes. It takes a big event to make people change their behavior. It is like getting people to wear seat belts or to stop littering. Dealing with climate change is this type of problem—getting people to support action to protect our planet. It will take a very big event to get the population on board, but when the time comes, it will be people like the safety and health professionals that will lead the way. Those of us trying to protect safety and health will get on board. Some earthshaking events will get everybody's attention. Safety professionals can complain about our human failing, lack of planning, and problems like infrastructure, but those upgrades to our roads and bridges may not happen until the bridge fails. It will cost so much more to replace all the roads and bridges after they fail. It is just human nature to wait until there is a disaster. In the movies, the aliens tell us to stop killing one another and fix the bridges, roads, and oceans, and maybe even get your kids vaccinated.

WHAT I LEARNED FROM BUSINESS

"Business is business!" These are some of the scariest words in my professional life. It implies that whatever a person or a corporation needs to do to make a profit is acceptable. To place the human factor behind a profit and productivity factor undermines any safety effort. The safety professional looks at the culture of a client, and if there is too much pressure on the business side, then there must be a balance added not just for safety but also for humanity to succeed with the safety and health program. The concept of balance is crucial. It is like the scales of justice in the law. There must be a balance between right and wrong in the law. There might be some bad decisions, but those bad ones will be balanced out by the good ones. The balance on the scale is then determined to be justice—not perfect but a good system that comes out somewhat equal in the end.

The same is true for safety. There is no money for the safety and health program when there are no profits. Some of the most dangerous industries are the ones that are losing money. Companies that are profitable have the funds to hire talented safety and health professionals to enhance their efforts. However, no matter how profitable a company is, there is always pressure to please the stockholders and to cut the fat of the organization, which can affect the safety and health effort. This is why balance is so important.

Sooner or later, middle safety managers feel pressure to push for more production and profits to enhance their upward mobility. That pressure for production and advancement must be balanced with the rewards of accomplishments in the safety and health effort. There must be a balance in the workplace between making widgets, profits, and human needs. This is the marriage of the safety and health program to the corporate culture. This is not an easy marriage, because many of the rewards are on the business side and few are on the safety side. The upward mobility for the safety side must be just as lucrative as the productivity and profit side. There should be safety and health people on the board of directors, and safety people must make wages similar to the top productivity people. This concept has not been fully accepted within the industry, but there has been some progress. Companies are starting to realize that preventing injuries and illnesses is an important profit center and makes for good public relations.

Companies have made significant progress, and I have seen many with a vice president of loss control on the board of directors. Some safety managers are getting top salaries, and their safety programs reflect the effort of skilled personnel who are rewarded with salaries and prestige. They have a good balance.

SYMPHONY OF SAFETY AGAIN

I use this phrase all the time. I use it for controlling employers, to get them to believe that the construction and operation of a facility are not individual part but are parts that need to be controlled and conducted by a higher authority to make them work in harmony. The same concept of the symphony of safety applies to the corporate structure. There must be personnel at the top who are making sure that all the components are working together to make the music of the safety and health effort. The top management of the facility must conduct the symphony of safety. Operations and activities must be planned and organized in a way that safety, health, and human factors are considered every step of the way. The conductors of safety and health have the responsibility of making certain that one person, operation, or department does not create hazards for the others. This is the same as the concept for the multiemployer work site. One party must step up for control and coordination. When people pass the buck to lower-level personnel, it is a recipe for disaster.

CANNABIS

Marijuana might be the biggest deal in safety and health. It is being legalized across the world and has entered the workplace in a big way. It is likely being used by a large percentage of people, including workers, and the effects are not well studied. There currently exists a drug-testing and punishing approach, which I believe is the wrong approach. We are going to need a human factor approach that accommodates the problems people have with the need to take drugs, both legal and illegal. Today, many workers know how to beat the system that really does nothing but hide the problem. Losing a good worker because they had cannabis at a party a week earlier is not a productive approach. We need to try to have a more honest approach to this gigantic problem. Many jobs are not compatible with drugs, prescription or not, and the people we put in those jobs must have a commitment to sobriety. A school bus driver and a police officer must be clean and sober. There may be other jobs where it might be acceptable to have people mildly impaired. We need to study this area extensively to avoid a catastrophe. Various derivatives of cannabinoids are going to be on the market, and some will have more effects on performance than others, and this must be studied. At this writing, CBD, a nonpsychoactive ingredient in hemp, is sweeping the country. It has properties of relaxation, but the effect on the brain or human performance is unknown. We must determine the effects of all these new substances. This is precisely the reason that NIOSH was established along with the OSHA Act. NIOSH must be funded and supported to conduct these studies.

FIREWORKS

Let's play with bombs. Here is a finger-blowing, skin-burning activity that scares the animals but not the people. This is proof that animals might be smarter than people. As a child, I thought the fireworks were magic, and I can still see the magic in those dazzling displays. Still, as an adult and a safetyman, I also see the potential for danger even at the professional displays, and it seems wrong for people to put on their own display without any training or experience. My experience at amateur displays is that the people setting off the fireworks don't even know what to expect when they ignite what they have purchased. I have seen the explosions go the wrong way and have seen people and children get burned because of a lack of basic safety precautions. Still, it seems like an enormous problem when anyone at all can purchase big-time fireworks and do whatever they wish with bombs and mortars.

Fireworks, in some ways, represent the essence of freedom. We set them off on the Fourth of July to let everyone know how proud and patriotic we are, and this is also a time when families get together for an activity of mutual enjoyment. When it's done over the water with a professional crew, I can appreciate the entire experience. When amateurs think they can do the work of trained professionals, it can be serious trouble. These amateurs have not had the experience and training to work with fireworks safely. This fact, plus the fact that a lot of money can be made from fireworks, sets into motion a path to potential disaster. I have seen people storing fireworks under the heat of the sun or next to a fire or barbecue. I have seen people choose to ignite the fireworks with matches in a dry wooded area. The worst is when the firework is ignited and fails to fly into the sky, and the untrained person refuses to douse it in water. The amateur fireworks person doesn't want to waste the money they paid for the mortar, and when they try to light it again, it blows up in their face, and they end up in the hospital or worse. I don't want to be a chauvinist, but I must comment on the fact that I can never remember a woman being engaged in amateur fireworks. This seems to be more of a male thing. I frown on this dangerous practice, but it doesn't change the fact that every Fourth of July, I go to somebody's house and see close calls and small burns. I often ask if there is a fire extinguisher around the house because I am a safetyman.

SAFETYMAN DRAMA

The situation with the Fourth of July fireworks makes me think about the drama in safety, and being the safetyman, I am always spoiling the fun. The reaction is the same. It annoys people when I ask for the fire extinguisher or ask how deep the water is. Recently, I saw parents allowing their small children to play in a hot tub and a pool with no lifeguards or supervision. I ended up watching their children while they were having their own fun. This human reaction against caution and safety is fascinating. What is it that makes people deplore safety and caution? This is the drama the safetyman faces every day. It is mostly about preventing human suffering and sometimes about life and death, but people act like it's not important, and they get angry at me. They think I am trying to tell them what to do, but I am just trying to prevent suffering and pain. Wherever I try to intervene, even if they are doing something silly, like pouring gasoline on a campfire, they get really mad. They think they know more about these things than the safetyman, or they just don't appreciate the danger of their actions.

If I start looking for violations—for example, if I tell them their guardrail is too low for compliance with building standards or the electrical wire or extension cord they are using outside is rated only for indoor use—they can get really angry. It is like they are saying to me that everybody knows these safety standards are stupid. Worse, they think I am stupid. Most people roll their eyes, shake their heads, and walk away from the safetyman, who is only trying to protect their lives or their children. It is like they think safety is dumb. To me, being serious about safety is doing something important. I think I am giving them a free safety lesson or inspection, as I normally get paid to do safety inspections and training. Still, they don't care. The drama in safety is never-ending, and it is due to the attitude of many people that it is not important, even when their lives depend upon it. My work requires me to go where somebody else was hurt or killed so everybody else can be safe.

My experience allows me to evaluate and eliminate dangerous situations. But all that experience and training is useless when people don't listen or get angry. I take my own precautions (that's why I am still alive), but the main objective of the investigations is to prevent a reoccurrence and protect other people from the hazard. There have been times where I found a hazard that was unrelated to the investigation, and I refused to move until it was corrected. Many people can't understand the importance of me standing down until the hazard is corrected. They are not used to someone representing safety. It has gotten to the point where I have told management that I will call OSHA. I have called OSHA a few times. In some investigations, I found evidence that was removed or put out of sight. Sometimes, people lie to me about what happened! This is all part of the drama of being a safetyman. It is important for me to keep control of the investigation, because if the safetyman loses control, it may not turn out too well. When I show my independence, it seems odd to many of these people who are used to getting their way. It always gives me an opportunity to be successful in my efforts to convince everyone about my sincerity about safety and health. I will even turn on a client that is paying my fee if they expect me to do or say anything that doesn't seem right or consistent with my beliefs. This always works to my benefit.

CORPORATE AMERICA

The safetyman is a businessman and supports corporate America. Corporations can and will do whatever is needed to support the safety and health effort. I can think of so many devices that have been financed and developed by corporations that have had tremendous benefits. Respirators, heart defibrillators, body cameras, testing devices, heat sensors, fire detectors, and smoke detectors have advanced the safety and health profession. Measuring devices that used to cost thousands today are much less, and many, such as sound-level meters and light meters, come as applications on my cell phone. The possibilities for corporations within American and outside American for advancement and improvement of the quality of life for humans all around the world are amazing.

Despite the overwhelming evidence of the benefits from corporate culture, there are still too many situations where some corporations sacrifice safety and health for profits. The safety culture expects objective evaluation, measurement, and the necessary investigation and reporting of defective products and materials, and sometimes the organizations that are not standing up to improve operations and products in the face of injuries and illnesses. In a few organizations, it is a standard operating procedure (SOP) to deny responsibility and liability for hazards associated with their operations and products. Cases like the BP Deepwater Horizon oil spill that occurred on April 20, 2010, considered to be the largest marine spill in the Gulf of Mexico, are indicative of the problem. A White House commission blamed BP and its partners for a series of cost-cutting decisions and an inadequate safety system. There are reports from 2012 indicating that the well is still leaking. On the other hand, as a result of all the bad publicity and litigation, BP is now a leading corporate advocate for safety and health.

Since the 1950s, people have been suing cigarette manufacturers for marketing and deceptive advertising, and there is overwhelming evidence that cigarette companies failed to do the right thing to protect human life. Another example is the issue of safe drinking water. The people of Flint, Michigan, were poisoned by their own government when, in 2014, the drinking water source for the city was changed from Lake Huron and the Detroit River to the cheaper Flint River. Studies determined toxic lead leached from water pipes into the drinking water for more than one hundred thousand people. In 1962, Ralph Nader told us the Corvair car being sold by General Motors was "unsafe at any speed." This list is long and doesn't include thousands of chemicals and drugs put into the stream of commerce that have been determined to be harmful. Organizations and industries have done the wrong thing for profits and failed to admit their guilt or make it right. Sometimes these organizations spend money to avoid making changes to their products or procedures. Recently, Johnson & Johnson was alleged to have used asbestos in their baby powder. Johnson & Johnson states that these allegations are false, and they use only pure talc that contains no asbestos or any harmful material. The American Cancer Society, on December 4, 2018, indicated that in its natural form, some talc contains asbestos. There is a history of many organizations focused only on profits and losses, and the safety and health of people have not been a primary consideration. There is a real question about whether our society puts too much pressure on profits and not enough on humanity. It's great to make the money, but we also must take care of people.

QUALITY SAFETY PRODUCTS

There is a scandal about hearing protection right now. Allegedly, the hearing protection sold to our military didn't do the job as promised.

Safetyman knows that the noise is a variable, the ear canal is a variable, and even the health and hearing of each person is a variable. It is a problem for safety equipment and safety products to predict effectiveness under all conditions. Respirators are one of the most important types of personal protective equipment because contaminated air, and sometimes too little air, will end or shorten your life. The most important aspect of respiratory protection is called the quantitative "fit test." This is a proven scientific methodology that can be done with instruments that are inserted inside the respirator that monitor the air and potential contaminants and assure the high level of protection. Of course, there are variables even in this case. This fit test is useless with my beard. It is only a good test when there is a good seal on the respirator.

There are some other tests called qualitative fit tests, which are not as effective as the quantitative fit test. With these tests, some strong-smelling oil is used around the respirator, usually banana oil, and the human is asked if they can smell the oil. If the answer is no, then it is accepted as a good-fitting respirator. I personally choose the quantitative fit test. I want to know exactly how much bad stuff I am breathing. None of these tests will result in 100 percent protection for the user. If a user wants anywhere near 100 percent protection, they are going to have to get into an encapsulated suit with supplied air. This is the kind of protection needed for asbestos or Ebola. The science and the choice of these products for personal protection and safety products is an important aspect of being a safety and health professional.

Hard hats are not all the same and don't all fit the same. Never use a hard hat without a chin strap. We hardly ever see workers with chin straps. Hockey and football players have chin straps for a reason. Today, concussions are viewed as a much more serious injury than in previous years. Fall protection with body harnesses, lanyards, and anchorages involve different hardware and software with different capacities. It is important to check with the manufacturer or distributor of the fall-protection equipment and systems before using or recommending them. Even high-visibility vests vary from manufacturer to manufacturer as to color and illumination. People exposed to vehicular traffic need the brightest easy-to-see materials. These can be life-and-death decisions. The good safety products and equipment are not the cheapest. Wrenches, drill bits, and any component might be substandard or a knockoff of a legitimate product. Knockoffs of products and materials have caused many injuries and deaths. There can be a big profit in knockoff tools or components. On one of the big projects in Chicago, it was discovered that the electrical switches were knockoffs and didn't meet the project specifications. Crane and equipment failures have occurred where the component that failed was not original manufacturer's equipment (OEM). Knockoff wrenches and substandard drill bits broke and flew across the room. Abrasive wheels on grinders fly apart and injure people like a shotgun blast without proper protection. Every safety professional must stay up to date with the changing technology. One of the biggest benefits of attending safety conventions is seeing the displays of the latest technology and products that improve safety.

SCIENCE-FICTION ROBOTS AND COMPUTERS

Many safety and health advancements have come right out of science fiction. Simulators to test aircraft and other machinery, robots, and machinery come right out of the pages of those books and television shows. We are close to self-driving cars and trucks. Much of the medical and testing equipment utilized today were once visions of the future. A safety professional must keep an eye on both science and science fiction to stay in touch with the future. We are behind in the use of satellites and GPS in tracking the location of people on the job or in hazardous locations. Robots are used to perform surgery, and enclosed and encapsulated environments used to manipulate and clean hazardous products, such as sandblasting operations, were otherworldly at one time.

Many innovations and solutions to problems arrive from science fiction. We cannot ignore looking into the technology used in every field for application to the safety and health of our people. A safety professional must look at technology to see what might be available for the protection of people. Virtual reality and artificial intelligence offer the possibility of having protection that goes beyond a person's own eyes and vision. Applications of virtual reality to safety and health offer so many possibilities that it is hard to know where to start looking. Glasses that give people vision in all directions can prevent many slips, trips, and falls. Programs can be connected to visual aids that will pick out hazards before they can cause injury or death.

Science fiction today is close to science fact. The solution to safety and health problems may be right in the palm of our hands, using technology, especially robots and computers. These robots and computers will aid the human in our capacity to inspect, evaluate, and avoid dangers. They won't have injuries and illnesses, and they won't make mistakes. There will be less need for standards or regulations because robots will enhance human ability. Robots and computers will do jobs that humans should never do, ones where they are exposed to deadly health hazards. After the massive March 11, 2011, Tohoku earthquake, subsequent tsunami, and radiation leaks at the Fukushima nuclear power plant, which continues until this day, they initially used humans for an attempted shutdown and cleanup, and the result was thousands of deaths and uncountable human suffering. This cleanup could be done safely only by robots. Today, robots and computers are being used for the cleanup, and without these sophisticated robots, there would likely be no cleanup at all, and the earth might be in significant peril. These robots resided only in the minds of science-fiction writers and on the pages of science-fiction novels, yet now they are working to save our people and our planet. These robots and computers will not get sick from radiation, and they won't hurry the job to increase production or take a chance to keep the line moving.

The current space program, both public and private, would be impossible without the use of robots and sophisticated computers. The robots will eventually solve many safety problems with which we are now confronted, and there will be bigger contributions toward climate change, fixing our environment and our movement into outer space. It will no longer be considered science fiction at all. Some people believe jobs will be created when humans program and maintain robots and computers, but those jobs might go to robots and computers too. I tell my hairstylist that she will be replaced with a helmet computer one day. I want the exact same haircut every time. If a computer cuts my hair once the way I like it, I don't see why that helmet can't be programmed to do it again for the rest of my life. This could happen to many jobs. In medicine, many surgeries are only performed by robots that have perfectly steady hands. I fear humans might get lazy and careless.

Robots and computers are great assets for safety and health. They don't have the many human characteristics that make us vulnerable to injuries and illnesses. They are machines without feeling and without human needs. They don't need vacations, and they don't go to the bathroom. They don't get angry or moody. They don't make many mistakes. They don't get sick. When you come to be a safetyman and you are trying to eliminate errors and mistakes, robots and computers solve problems that you see every day. Robots will control the cameras and virtual reality that we need to watch people and watch out for people. Think about cameras with brains, reasoning, and artificial intelligence. Robots and computers will take over the world and take over for the safetyman too. The robots and computers will see how many mistakes we make and how we fail toward perfection.

Isaac Asimov wrote about the advantage of robots and computers over humans and how their superiority will win out because we are so flawed with our primitive impulses. Eventually, in science fiction, robots can no longer tolerate flawed humans, and they must be eliminated. Robots and computers will be programmed to eliminate our mistakes and accidents, and they will want to control and perhaps eliminate us. This fear of being controlled by robots or even eliminated exists in the back of our minds. Asimov's imaginary robots were programmed with a directive that they could not harm any human, but somehow, these robots just couldn't be controlled, and war broke out, and robots were winning until human creativity and thinking outside the box eventually prevailed. The safety professional of the future may be a robot able to eliminate accidents and injuries before they even happen.

DRONES

Drones are a type of flying robot we are already using in the safety and health profession. There are places I have never been able to go and places it is much safer to send a drone to. Drones are used increasingly in the industry to prevent accidents and injuries where workers have difficult access to great heights. They come in handy for inspecting wind towers, communication towers, bridging, and other difficult-to-reach locations. These drones are leading the way in showing us how technology will be used in the future to prevent accidents and injuries. Every time FedEx or UPS, and now Amazon Prime, comes to my home or office to deliver packages, I realize how much better and safer it will be when they have drones or robots make these deliveries. I notice that too often the human drivers are in a hurry. It's like they are in a race or on an incentive program to get the most deliveries in the shortest amount of time. It is amazing how many of these people, trucks, and cars we see every day. It is getting out of hand with the push for one-day delivery and people not wanting to leave their homes to get food, drugs, or anything.

The biggest changes are needed in the world of technology. The potential of using robots, computers, and artificial intelligence to eliminate danger is extraordinary. A friend of mine who operates drones for a business needs to practice with his drone on the weekend. He goes to a local park and operates his drone. He has been going to the park on weekends for several years. He told me last weekend his drone was contacted by satellite, and the satellite told his drone he was in the restricted territory. Several days later, he got a letter from the Federal Aviation Administration. Here was all the new technology in action, and the satellites were monitoring and communicating with the drones.

ACCEPTABLE RISK CONCEPT

Until we get the robots and the computers to protect our world, there are so many problems that can be solved if we have enough safety men and women to apply what we have available today. It's a question of our self-preservation and quality of life, and it must be our highest priority. Using available technology to make the world safer is not a high priority. We can look at our failing infrastructure as an example. We cannot wait for our bridges and roads to crumble before we take decisive action. As safety professionals, we must believe being proactive for the protection of our health and safety is the highest priority. It makes sense that the protection of our climate and environment would also be in that category. We must develop strategies for our own protection and advocate for improvement. We need to live in the world as informed people armed with the most modern technology, information, and training to survive and improve dangerous and unhealthy conditions. Safety and health have deteriorated over many years. Just look at the situation with obesity. We must have more information and make better choices. There will never be a perfectly safe or healthy world, but I believe we need to put the days of acceptable risk behind us. The idea of acceptable risk is still prominent in our world.

Advocates say that if you are an astronaut, you accept the risk of the space shuttle exploding, or if you are a skier, you accept the risk of falling or being hit by another skier. Hang gliding, bungee jumping, and mountain climbing are other examples. Some people believe that when you go out in the world for any reason, you are accepting the risk. They say this without calculating the risk. When you are a safety professional, you investigate the actual facts of potential injuries or incidents and determine the risk. We may also determine if there are responsible parties acting in the best interest of the people exposed to the risk.

Taking the examples from above, the astronauts are not at fault if the space capsule was improperly designed or somebody pulled the wrong lever or failed to act in a reasonable manner. Skiers don't have control of avalanches, the number of skiers allowed on the ski hill, or the type of conditions and equipment. Skiers might not know that skiers and snowboarders will be skiing in the same location. Hang gliders, bungee jumpers, and mountain climbers can't be responsible for defective equipment. The problem with the concept of acceptable risk is the risk is placed on a person who is relying on other parties who have represented that they can engage in an exciting activity without serious injury or death.

It might be that the astronauts, skiers, bungee jumpers, and mountain climbers took an unnecessary risk that caused the loss, but we have to investigate the cause. If a carabiner failed to hold the rated load and caused the fall, then it is wrong to blame the mountain climber for their injury or death. No matter what activity we engage in, we depend on other people to make it safe and sound. To me, as a safetyman, any of these activities are not much different from riding in an elevator. That elevator must be safe, and other people were responsible for my safety when they constructed, tested, and maintained the elevator. Unless I do something highly unusual to risk my life, those elevator specialists are responsible for my safety when they offer me the opportunity to ride in the elevator. There is some risk, but it is not an acceptable risk.

Is it acceptable to drink coffee and alcohol and smoke cigarettes? Each of us must make choices each day. We need to make informed choices. When we take illegal or unregulated drugs, we take an unacceptable risk, but if we are using products that are available, regulated, and put in the stream of commerce, a safetyman believes that we should be able to feel that those products will be safe and will not injure our health or shorten our lives. This is a safety issue that has been litigated many times. Thousands of people have been compensated for asbestos exposure and cigarette exposure, including secondhand smoke, and there have been numerous cases about chemical products and by-products. The product manufacturers and companies that manage human activities have an obligation to assure, under normal conditions, their products and activities are safe and healthful. If an individual does something unusual, illegal, or unexpected, that is a different situation entirely. When an individual knowingly or intentionally does something dangerous, then even the safetyman will say they took an unacceptable risk. This whole process is complicated and constantly changing; however, for the purposes of this book and from the perspective of the safetyman, I believe our society has a responsibility to be protective of people conducting activities under all circumstances unless they exhibit some aberrant behavior.

GIVING UP ON SAFETY

Sometimes it seems hopeless—the whole idea of changing human attitude and behavior. People like to take chances, and they like action. People want the freedom to act out and do dangerous things. The entire society seems attracted to danger. I think of television shows like the one where they try to get the crabs in Alaska or even some of the most popular shows like *America's Got Talent*, where, often, the acts that go to the next round are the ones where people are risking their lives. There are the Flying Wallendas and so many other examples of this problem where we are trying to protect people who may not want to be protected at all. The trouble with this is seeing a victim who has lost a hand in a machine or has fallen from heights, and their lives are changed forever, and their families are severely affected. Society is affected, and even though a safety professional might sometimes get discouraged, we can never give up the fight because even if just one person is saved from a preventable injury or illness, the work is worthwhile. There are times when after so many years, I get discouraged and feel the momentous weight of human nature fighting against all my efforts, but when I make a change or influence even a single person or institution to establish safety and health, I know that I may have made a difference in this world—and maybe even a big difference over the years.

A WORD TO FUTURE SAFETYMAN AND SAFETY WOMAN

It is exciting to think of the difference a safetyman and safety woman can make in this world. It is an emerging profession, and very few other people are looking at these problems. People are busy doing many other things in this world, especially in the areas of technology, but it is only the safety professional who is in a position to look at the big picture and integrate the progress made in technology into protecting our quality of life. There can be no more rewarding career than being a safety professional. It is not easy to get the attention of the rest of the people who think it is not important until something bad happens to them or their family, but the people who are saving the lives, saving the bodies, and saving the children will eventually be rewarded and appreciated. It can be a frustrating fight, but in the end, the saving of life and health will be appreciated, not to mention saving all that wasted money.

As a person with this experience, I believe the safety people who will change our culture must be vigorously independent, objective scientists. The situation in our society today, where everyone is on one side or the other, makes getting consensus and cooperation difficult to achieve. Those of us who do achieve an independent status will be the most valuable people on earth. We will be the problem solvers, the integrators, and the people who both sides will turn to when they want to live a safe and healthful life. We must have compassion for people and have a good work ethic. I believe my strength was my joy of working hard every day and feeling like I was doing something important.

One day, people from all sides of the culture will appreciate efforts to make our world safer and more healthful. These people will turn their attention from their political battles and squabbles, and they will be united in our effort to make the world safe and healthful. If I had to do it all over, I should have been more collaborative, trying to get people who don't understand the importance of safety and health on board. I could have spent more time recruiting an army of people like me.

SOME PERSONAL STUFF

A VILLAGE OF SAFETY

I HAD GOOD EXAMPLES OF BEING A SAFETYMAN. MY FATHER was a moral person. He taught me that the world could be cold. He lived in a cold, hard world during the Depression. My father believed in working hard. My mother was an amazing businesswoman. She could have been the head of a major corporation. She ran the family like a clock. You could never get away with any plan to fool her, and she talked to so many people every day that she was an amazing source of information. I would say my mother was like the internet before there was an internet.

Figuring out what to do with my life was hard. I realized my father was working so hard to provide for our family by being a salesman. It seemed unrewarding to me just to be selling a product. I wanted to contribute and make the world a better place. My mother was what we used to call a homemaker, and she was a very capable and positive person. Without her positive affirmation toward me, I could never have become a safetyman. As I grew into adulthood, both my parents could see me as the head of my own safety and health consulting firm, and they supported my goals and dreams in every way possible. My uncle was a well-known and successful car dealer, and he passed on to me the love of cars. I still love cars. Figuring out how to make a living seemed like a difficult problem. My parents insisted all their children have jobs to make money at an early age. They wouldn't give us money for dating and fun if we were not working. I am grateful to my parents for giving me and my brothers such a great work ethic. It has been good for all of us and has kept us out of trouble.

FAMILY VALUES

We had family values of hard work. We never expected anything for nothing, and we knew that if we wanted to achieve some success, we could do it by working at it every day. We were tough, and we were taught that we had to work hard or we would end up in trouble. The values were conservative in that we learned not to take chances with our future. There was not too much praise and maybe a little too much criticism. I learned at an early age from both my parents that life was not going to be easy and that hard work was the best measure of my life. Both of my parents helped me to realize that failure was the road to success. Whenever I was criticized, I learned something. The compliments, on the other hand, I found to be more self-serving and not at all accurate.

When I see children today never getting criticized or even told the truth, I wonder how they can become successful. I like criticism and failure. I learn from my mistakes, and I try not to make the same mistakes. I learned from my family and life that it is the hard things that matter. Some of the things that come easily to me, like talking or presenting, are not always the more important aspects of my work.

It's the hard stuff that gives me the most satisfaction, like trying to write this book. When I first started out as a safetyman at OSHA, my supervisors made me spend at least three months memorizing the OSHA Act and the OSHA standards. I was thinking the whole time that it was the most boring activity on earth, and I just hated it. I had no choice, so I memorized the standards every day. To this day, I have most of those standards memorized. It helps me with my work in so many ways. It was the same with the writing of standards and working and meeting with people with so many different objectives and points of view. Getting standards passed was going to take years and compromise and patience, and it was hard. But I knew after the work would come the rewards. I know the standards we wrote by heart, and I developed professional relationships and experiences that would benefit my career and my profession.

CONCLUSION

THE FOUR PS

THERE ARE PROFITS, PRODUCTION, PROMOTIONS, AND PUBLIC relations. These are the most powerful incentives for workers and all humans. To succeed in an organization, people must serve other individuals who are focused on the four Ps. To interrupt or interfere with these Ps is a courageous act. When safetymen and safety women say *stop* or *slow down* or even ask for deeper analysis, they can be considered to be causing waves, and they risk being fired or languishing in lower-level positions their entire career. This heroic dedication is necessary for the safety people to balance these forces, and therefore, it is difficult to get safety integrated into our world. People remember so many incidences where safety was not on the front burner, from the *Challenger* disaster to the Deep Horizon spill and calamities all over the world, including Chernobyl. The reason is the four Ps.

SAFETYMAN SONG

When I was a trainer at OSHA National Training Institute, I realized the power of the four Ps, and I wanted to teach the safety and health people who were out training the industry and the OSHA compliance officers how difficult it is to fight against the forces of the pressures for production and upward mobility. I knew the power of a demonstration and audience participation, so I devised the "Safetyman Song." This song was from the old *Zorro* television series, but I changed the words to the following:

> Out of the night when full moon is bright
> Comes a hero known as Safetyman.
> He is bold and brave.
> Your life he will save.
> He is a hero know as Safetyman.
> If it is fire you fear,
> And you need turnout gear.
> Call a hero known as Safetyman.
> If you think you might fall,
> There is a man you can call.
> He is a hero known as Safetyman.
> (Then the chorus and more verses.)

This song was supposed to be a powerful tool to get safety people to commit to safety. The idea was they were dedicated and would stick to the ideals of being a hero for the safety and health of other people.

I know it had a positive effect because it is the single most remembered thing from the hundreds of training courses I have conducted. One time, when I was in another country, I was walking around, and I heard somebody sing the "Safetyman Song."

On the other hand, I have been ridiculed for this song. Some people say it is clownish or that it makes light of a serious topic. It just goes to show that when you do something, you bring attention to yourself. I am proud of the "Safetyman Song." I think it is one of the most important things I have done.

HOWIE

Howie is my dog. Every morning, for more than sixteen years, he heads to my office to sit with me and help me do my work. Every day with him is a treasure. He is old now. He has lived almost two years longer than predicted by the veterinarian. Howie has a cancerous tumor, but he doesn't let it stop him from living his life. Howie is such

Howie's sixteenth birthday

an inspiration to me of what it's like to age and become infirm. He is so strong and fights for his life. I keep learning more and more from Howie, and some of it is about safety and being a safetyman. Howie, in many ways, is smarter than humans. I know everybody thinks that humans are smarter than dogs, and there is even a humane group that refers to dogs as our dumb friends. But Howie is not dumb at all. Howie is very cautious in his movements and activities. He never takes chances, and he understands being careless might cause an injury. The only time Howie ever got hurt was when he was young. He was dropped on the stairs and tore his ligament. It had to be surgically repaired and has lasted all these years. Howie is a safe dog and is mostly dedicated to his work with me. He has absolutely no meanness in him even though we know, at his advanced age, he suffers from many afflictions. In these last few months, he has mostly slept, and we must carry him up and down the stairs. He is so good. He is so loving, and he gives me a reason to go on in this terrible, dangerous world every day. If Howie can make it another day, so can we.

GOLFING SAFETY STORIES FROM THE SAFETYMAN

SPORTING ACTIVITIES ARE ASSOCIATED WITH GREAT DANGER, and there has been reluctance to make upgrades and improvements. The ASTM F08 committee, where I have worked with excellent safety people, has been working toward improving the condition for the players and the fans. I have been to many sporting events and have seen players and fans getting hit with the puck or the ball. There is a lack of guardrail and netting protection for the fans. There is also not enough training. Most recently in Houston, at the World Series, I saw a foul ball hit a glass partition and shatter it, sending some shards of glass into the seats.

My own experience in sporting danger has revolved around golf. From my earliest memories, my dad wanted me to play golf and be a good golfer. Dad was a pretty good golfer, and he so wanted to play with me and my brothers. My first job was being a caddie. I was a teenager, and right from the start, I began to see the danger and perils of golf. There was the ergonomic problem of trying to carry two heavy bags of clubs, weighing at least forty pounds each, on my shoulders for eighteen holes. This load was completely out of balance and would leave me with pain and friction burns for a week, and it possibly did some permanent damage. Even after suffering though those rounds, those county club golfers rarely tipped me for my effort. I must admit I was a lousy caddie. I had trouble tracking the flying golf balls. This is an important human factor, which makes golf so dangerous. Humans, not just caddies, are limited in their ability to track spatial relationships, especially with a flying object. It is hard enough for a human to track the location of an overhead power line, let alone a flying object. The country club members were never happy when I could not find their ball.

Golfers are always trying to play their rounds fast because the round normally takes four hours or longer. They hit their balls into the group ahead of them on the course. They get angry when the group in front is playing too slowly. The rule as a caddie was to yell "Fore!" when the ball went flying toward the people playing in the group ahead. This "Fore!" is supposed to warn people to run out of the way and may be the worst safety plan of all times. That ball travels in the air at high speed and is as hard as a rock. "Fore!" is just not that helpful.

I remember traveling to Las Vegas with my brothers to play golf, and I had the opportunity to take lessons every morning. I kept thinking that if I had more lessons, I would get better and be able to control that little white ball like the professionals. The lessons did not do much good. I had developed a friendship with the professional golfer at the Dessert Inn, and he knew I was not getting much better at golf, despite the lessons. My lack of improvement did not stop the professional golfer from inviting me to play in a tournament for charity. This was a tournament called a scramble, where they would only use the best shot from the group. This idea of using only the best shot was comforting to me, as my poor play would not stand out to the other players.

When I arrived at the course, I was surprised to find my name on the leader board. I was playing with the top golfers in the area, and we were rated number one to win the tournament. I was nervous, and my first two shots off the first tee sliced into the parking lot, and one of them hit a car. Well, at least this was only property damage. My poor play caused my whole group to play slowly, and besides having the rangers telling us to speed up, the group behind us kept hitting us with their golf balls. In this case, nobody was seriously injured. but on other occasions, I have seen people hit in the head with the ball and carted off the course. At the end of the round, our group finished second in the tournament, and by that time, some of the other golfers in the group were not too happy with me and my poor play. One of the professionals told me that I had cost him some money.

After my shower, I went to see the professional at the Dessert Inn and asked him why he had put me in the group with the best players on the course. He took me over to his closet and presented me with a brand-new set of Ben Hogan golf clubs as a thank you for causing his buddies to lose the tournament. He had used me as a cooler.

One of the other problems I have found with the dangers of playing golf is the human emotion combined with the driving of golf carts. I have seen golfers speeding and going at an angle that can tip over the cart and even hit obstacles on the course. There should be a training course for these carts.

Flying objects, lack of training, and human emotion make for a deadly combination. I have seen golfers break their clubs and throw them. I have seen clubs thrown so often I must believe it is not uncommon. Golfers get so angry at their game or at themselves that they drive the cart into the water or into a tree. On two separate occasions, I have seen a bird injured or killed by a golf ball. Sometimes, golfers will drink too much, and one time I played with a friend who was so drunk he ran the cart into a railroad tie at high speed.

There was a newspaper columnist by the name of Mike Royko in Chicago who used to write about golf. He had a column about claims that golf equipment manufactures made that they could improve the average golfer's score. He had found that such claims were not accurate, and I can tell the reader that I agree. In my experience as a safetyman trying to play golf, I have found that the average golfer, no matter what equipment or accessories, has little idea of the direction of the ball. I have seen these golfers hit the ball ninety degrees to the right and left and, on occasion, backward over their heads. I have seen and even experienced having the head of my club break off from the shaft and go farther than the ball. I have seen golf cart collisions, turnovers, and damages. Mixing all levels of skills and abilities with a hard ball, tapered club, and vehicles is a rough combination. Now just add some human anger, frustration, and a few drinks, and I think it explains everything about safety. A reasonable safety program for golf would start with personal protective equipment, separation of the pedestrians from the vehicles, and training for both the hitters and drivers. Do you think it might ruin golf?

www.ingramcontent.com/pod-product-compliance
Lightning Source LLC
Chambersburg PA
CBHW031816170526
45157CB00001B/81